Sources and Studies in the History of Mathematics and Physical Sciences

Sources and Studies in the History of Mathematics and Physical Sciences was inaugurated as two series in 1975 with the publication in Studies of Otto Neugebauer's seminal three-volume History of Ancient Mathematical Astronomy, which remains the central history of the subject. This publication was followed the next year in Sources by Gerald Toomer's transcription, translation (from the Arabic), and commentary of Diocles on Burning Mirrors. The two series were eventually amalgamated under a single editorial board led originally by Martin Klein (d. 2009) and Gerald Toomer, respectively two of the foremost historians of modern and ancient physical science. The goal of the joint series, as of its two predecessors, is to publish probing histories and thorough editions of technical developments in mathematics and physics, broadly construed. Its scope covers all relevant work from pre-classical antiquity through the last century, ranging from Babylonian mathematics to the scientific correspondence of H. A. Lorentz. Books in this series will interest scholars in the history of mathematics and physics, mathematicians, physicists, engineers, and anyone who seeks to understand the historical underpinnings of the modern physical sciences.

More information about this series at http://www.springer.com/series/4142

Dominique Raynaud

A Critical Edition of Ibn al-Haytham's *On the Shape of the Eclipse*

The First Experimental Study of the Camera Obscura

Springer

Dominique Raynaud
PPL
Université Grenoble Alpes
Grenoble Cedex 9
France

ISSN 2196-8810 ISSN 2196-8829 (electronic)
Sources and Studies in the History of Mathematics and Physical Sciences
ISBN 978-3-319-47990-3 ISBN 978-3-319-47991-0 (eBook)
DOI 10.1007/978-3-319-47991-0

Library of Congress Control Number: 2016954711

Printed on acid-free paper

This Springer imprint is published by Springer Nature
The registered company is Springer International Publishing AG
The registered company address is: Gewerbestrasse 11, 6330 Cham, Switzerland

Table of Contents

Acknowledgments

I have benefited from much advice during the course of this long research on Ibn al-Haytham's *On the Shape of the Eclipse.*

This work was initiated several years ago when I first got Naẓīf's book in my hands and suspected some discrepancies between Naẓīf and Wiedemann's readings of *On the Shape of the Eclipse.* I first collated and transcribed the manuscripts. At this stage, I benefited from the linguistic support of Fatma Fersi (University of Grenoble), Hamdi Mlika (University of Kairouan) and Elaheh Kheirandish (Harvard University). I also express gratitude to Ken Saito (Osaka Prefecture University), Gregg de Young (American University in Cairo), and A. Mark Smith (University of Missouri) for insightful comments on a previous article of mine, published in the *Archive for History of Exact Sciences*, that contributed much to the present work in establishing the stemma codicum of Ibn al-Haytham's treatise.

I am especially indebted to J. Lennart Berggren (Simon Fraser University), Sylvia E. Hunt (Laurentian University), Gaute Hareide (Volda High School), and Zheran Wang (Beijing University), who made valuable suggestions to improve my translation and commentary, and Hassan Tahiri (University of Lisbon), who kindly revised my transcription of the whole Arabic text, paying particular attention to its grammatical accuracy. He also helped decipher two barely legible marginalia in MS B. I thank him for his patience and dedication.

St. Pierre, May 2015

Introduction

This book aims at providing the first critical edition with translation and commentary of Ibn al-Haytham's *On the Shape of the Eclipse*, which is the first experimental study of the camera obscura.

The motivation to undertake this edition results from the very significance of Ibn al-Haytham's treatise in the history of optics:

1. *On the Shape of the Eclipse* strictly adheres to the experimental method—an extremely rare occurrence in the middle of the Middle Ages—that allowed Ibn al-Haytham to resolve two outstanding issues from at least Late Antiquity: "Why does the Sun penetrating through quadrilaterals form, not rectilinear shapes, but circles?" (*Problemata Physica* XV, 6) and "Why is it that during eclipses of the Sun, if one views them through a sieve or a leaf the rays are crescent-shaped in the direction of the Earth?" (XV, 11). As Ibn al-Haytham's solution is closely dependent on the use of the experimental method that inspires the whole treatise, I defer this discussion to a more suitable place (see Chapter 3, pp. 95–7);

2. *On the Shape of the Eclipse* provides the first successful attempt to merge the two branches of Ancient optics, which finally resulted in the abandonment of the extramission theory. This bestows historical significance on Ibn al-Haytham's work, since there is no prior history of this approach. In this regard, the treatise represents a key milestone in the history of optics (Section 1).

3. *On the Shape of the Eclipse* also includes pioneering research on the conditions of formation of images—in a time deemed to be committed to aniconism. However, it is an open question whether Ibn al-Haytham's work laid the basis for future investigation on the camera obscura (Section 2).

1. A Key Milestone in the History of Optics

Ancient and Medieval optics were divided into two special branches: *optics* proper (*optica, 'ilm al-manāẓir, de aspectibus*), centered on the study of sight and visual perception; *burning mirrors* (*catoptrica, 'ilm al-marāyā, de speculis comburentibus*), which focused on the geometric analysis of light, thereby laying the foundations of modern physical optics. One was the science of visual rays; the other was the science of luminous rays. Both parts have Greek roots. The first is found in Euclid's *Optics* and the second in Diocles' *Burning Mirrors.* Both works were acquired by scientists in medieval Islam.[1]

Ibn al-Haytham is generally credited with being the instigator of the unification of the two branches of optics (Sabra 1989, II: liv; Smith 2001: cxvi). More specifically, his intromission theory would have allowed him to break through the barrier between the science of visual perception (the direct vision, studied in Books I-III) and the science of light (the reflexion, studied in Books IV-VI). This unification was possible because of the symmetry of the laws of optics: the law of reflection is the same, whether the ray of light is emitted by the eye to the visible objects (extramission) or is communicated by the objects to the eye (intromission).

There are strong signs that *On the Shape of the Eclipse* was an early work by Ibn al-Haytham, which he composed when he began his optical masterpiece—probably soon after *Optics* I, 3, which is quoted in it—and that it already aimed at integrating the science of vision (physio-psychological optics) and the science of light (physical optics). In this work, indeed, Ibn al-Haytham does more than just apply geometrical optics to explain what causes an image to form in the camera obscura; he mostly questions the conditions of visibility of the image of the Sun and the Moon cast in

---

1. This division was perpetuated for another several centuries in the West. At the time of considering the problem of pinhole images, Roger Bacon chose to develop the subject in full in his *De speculis comburentibus*, which seems a strange choice. Regardless, it is worth noting that, while the projection of images through a screen does not require any mirror, the phenomenon involves radiations of light—not sight. This explains quite well why Bacon attributed this discussion to the science of burning mirrors, i.e., the science of light.

the darkroom. This approach is notably reflected by the phrases "perceived by the sense," "perceptible" and "imperceptible" frequently recurring in his work.

If one accepts that *On the Shape of the Eclipse* was one of the early works by Ibn al-Haytham—a fact supported by both astronomical dating (Appendix, pp. 161–86) and the rudimentary reference to Apollonius' *Conics*, of which he was soon to become a connoisseur—this treatise should be seen as the first accomplished work in which the two branches of Ancient optics were unified in one synthesis.

2. Ibn al-Haytham's Legacy

In a famous lecture to the French Academy of Sciences, François Arago (1839: 250–1) credited Giambattista della Porta with being the inventor of the camera obscura (*Magia Naturalis*, 1558). Two years later, Guglielmo Libri (1841, 4: 303–314) corrected Arago's error from excavating three texts prior to that of Della Porta: Girolamo Cardano (*De Subtilitate*, 1550); Don Papnutio (*Di Lucio Vitruvio de Architectura Libri X*, transl. Cesariano, 1521); Da Vinci, e.g., *Codex A*, fol. 20v (ca. 1490). Subsequently, Curtze (1901) drew attention to the camera obscura by Levi ben Gerson, around 1329/42. Later Pierre Duhem (1913: 505) discovered another text on the camera obscura by the astronomer Roger of Hereford, dated 1178. Then Wiedemann (1914) published his German translation of Ibn al-Haytham's work on the eclipse. In the late sixties, Lindberg (1968, 1970*ab*) reconstructed the centuries-long history of pinhole images (a tradition that only differs from the study of the camera obscura in that the device is not necessarily equipped with a screen). Since then other texts have been discovered, such as that of Guillaume de Saint-Cloud, around 1290. We now know that the result—not the operation—of the camera obscura was already described by the Chinese philosopher Mo Zi before 391 CE (Graham 1978: 375–9).

As a consequence, the question is no longer to decide who, among Della Porta, Maurolico or Da Vinci, invented the camera obscura, but to put each person in the right place in the long course of this history, and to precisely determine what followers owed to predecessors. As the camera obscura gained some kind of popularity in

16th-century Europe—to cite but a few: Reinhold (1542), Gemma Frisius (1545), Cardano (1550), Della Porta (1558), Barbaro (1568), Danti (1573), Benedetti (1585), Kepler (1604), Scheiner (1612), Schwenter (1636), Kircher (1646)—the question is also to know if the interest of the early modern scientists for the operation of this instrument owed something to the acquaintance of scholars with the medieval tradition of the camera obscura and, in particular, if some of them benefited Ibn al-Haytham's pioneering research on the subject.

In this respect, it is worth noting that some science historians consider such legacy unlikely (e.g., Lindberg 1968: 156) whereas others are inclined to believe it possible (e.g., Goldstein 1985: 141).

One major argument in favor of such legacy is that at least seven treatises by Ibn al-Haytham were available to subsequent scholars:

1. Apart from Latin and Italian translations, the *Optics* (*Kitāb al-Manāẓir*) was referred to ca. 1085 in the *Istikmāl* by al-Mu'taman Ibn Hūd, King of Saragossa, who cites the lemmas for solving Alhazen's problem discussed in Book 5 (Hogendijk 1986: 49), ca. 1230 by Jordanus de Nemore (*De triangulis* IV.20), who quotes "19 quinti perspective" (Clagett 1964: 668–9; 1984: 297–301), ca. 1250 by Bartholomaeus Anglicus (*De proprietatibus rerum* III.17) who cites the "auctor pespective" (Long 1979: 39–45).
2. *On the Rainbow and the Halo* (*Maqāla fī al-hāla wa-qaws quzah*) was referred to ca. 1170 by Averroes in his *Middle Commentary on Aristotle's Meteorology* where there is a mention of "Avenatan in tractatus famoso" (Sabra 1989, II: lxiv).
3. *On Parabolic Burning Mirrors* (*Maqāla fī al-marāya al-muḥriqa bi al-qutū*), translated in Latin by Gerard of Cremona, was known to Bacon (*De speculis comburentibus*), Witelo (*Perspectiva* IX, 39–44), the 13th-century *Speculi almukefi compositio*, Jean Fusoris (*De sectione mukefi*) and Regiomontanus (*Speculi almukefi compositio*).
4. *Book on the Completion of the Conics* (*Maqāla fī tamām kitāb al-makhruṭat*) was known to Maimonides who partly translated it around 1231 in his *Notes on Some of the Propositions of the Book of Conics* (Langermann 1984).
5. *Commentary on the Almagest* (*Maqāla fī ḥall shukūk fī kitāb al-Majisṭī*) was available to Judah ben Solomon ha-Cohen of Toledo, at the time when he composed his encyclopedia *Midrash ha-Ḥokhmah* in 1247 (Langermann 2000: 377).

6. *Commentary on the Premises of Euclid's Elements* (*Sharh muṣādarāt kitāb Uqlīdis*) whose Books V–VII, X and XI were translated into Hebrew by Moses ibn Tibbon of Marseilles, and Book X was given a another translation by Qalonymos b. Qalonymos of Arles (Lévy 1997: 434). This work was subsequently available to Levi ben Gerson when he wrote his *Commentary to Books I–V of the Elements* (Lévy 1992: 87, 90).
7. *Epistle on the Quadrature of the Circle* (*Risāla fī tarbi' al-dā'ira*) was quoted before 1350 by Meyashsher 'Aqov, alias Abner de Burgos (Langermann 1996: 50).[2]

It must be noted that none of these subsequent texts were known to us before special research was done to establish the facts. So the possibility that other tracts by Ibn al-Haytham were known to scholars—especially Hebrew savants who had some mastery of Arabic—should not be precluded *a priori.*

The most persuasive counter-argument against the survival of this work in later periods is the length of time that elapsed before the solution was rediscovered by Kepler. Notably, why did major perspectivists such as Bacon, Pecham and Witelo fail to understand the formation of images? Only two medieval authors, Egidius de Baisiu (Mancha 1989: 14) and Levi ben Gerson (Goldstein 1985: 48–49) approched the correct solution without, however, offering a fully satisfactory answer. After further intuitions by Da Vinci, the true explanation for the formation of the image was provided in Kepler's *Ad Vitellionem Paralipomena* (Kepler 1604: 48–56).[3]

---

2. This list sidesteps *On the Configuration of the World* (*Maqāla fī hay'at al-'ālam*), a work translated into Latin as *De configuratione mundi* under the auspices of Alfonso X el Sabio (*versión alfonsí,* before 1284), into Hebrew by Jacob ben Makhir ibn Tibbon of Montpellier as *Ma'amar bi-Tekunah* in 1275 and again by Solomon ibn Pater of Burgos in 1332. The work was subsequently known to Levi ben Gerson, who cited it in *The Wars of the Lord* (Lévy 1992: 86). The reason for leaving it aside is that the authorship of this work has been disputed by Rashed (1993: 490–1) who, on the basis of internal analysis, argues that this work is most certainly a work by Muḥammad the philosopher rather than by al-Ḥasan ibn al-Ḥasan the mathematician (see Chapter 1.2, pp. 14–5).

3. Kepler undertook this research to solve the problem posed by Tycho Brahe, who noted that the apparent diameter of the Moon seems to be reduced by about one-fifth during a partial eclipse of the Sun. Shortly after his meeting with Tycho, Kepler returned to Graz, where he observed a partial solar eclipse on July 10, 1600 by using a camera obscura. He then realized that the anomalous image was due solely to the size of the aperture of the camera obscura. He recorded his observation in his Eclipse Notebook for the year 1600, and expanded his correct explanation four years later, in *Ad Vitellionem Paralipomena* (Kepler 1600: 399–401; Kepler 1604: 48–56; Straker 1970, 1981; Plate 3.2).

Are those points enough for deciding for or against the acquaintance of medieval scholars with of Ibn al-Haytham's work? My position is that all the guesses that have been made so far—both pro and con—are pointless due to methodological weaknesses. Historical borrowings can be established with confidence only through textual parallels (Raynaud 2009, 2012*ab*). Textual parallels can be discovered only when one has at his disposal an accurate source-text. It is hard to take a strong view about Ibn al-Haytham's legacy until a reliable critical edition is made. A major aim of the present critical edition is to prepare future research on the putative legacy of *On the Shape of the Eclipse* in Latin Europe.

Chapter 1

This Edition

1. The Need for a Critical Edition

Ibn al-Haytham, born in Baṣra in the mid-tenth century and died in Cairo after A.H. 432/1040 (Sabra 1989, II: xix-xxiv; Rashed 1993: 1–19), is the author of over one hundred treatises dedicated to the mathematical sciences, among which is the *Epistle on the Shape of the Eclipse* (*Maqāla fī ṣūrat al-kusūf*). Even though Ibn al-Haytham's work was written around the first millenium, the work only benefited three studies. These are: a free abridged translation by the German physicist Eilhard Wiedemann (1914), an extensive commentary by the Egyptian engineer Muṣṭafā Naẓīf (1942), and a short comment by the Egyptian-American historian of science Abdelhamid I. Sabra (1989).

Let us review them one by one.

1. Wiedemann's Translation

Ibn al-Haytham's *On the Shape of the Eclipse* was freely translated into German by Eilhard Wiedemann (1914: 155–169). As this translation was the only testimony on Ibn al-Haytham's work until that date, Wiedemann provided a valuable service to the history of optics. Wiedemann's transalation was made from MSS O and L. He was unaware of the existence of MS P, which is the more faithful manuscript. None of MSS O and L is complete. Both lack 14 words at lines 247–8 (time 0.344),[4] 56 words at lines 562–6 (time 0.789) and 83 words at lines 569–77 (time 0.795).

---

4. 0 denotes the beginning of the text, 1 the end of the text. See Section 2.1, p. 17.

D. Raynaud, *A Critical Edition of Ibn al-Haytham's On the Shape of the Eclipse*,
Sources and Studies in the History of Mathematics and Physical Sciences,
DOI 10.1007/978-3-319-47991-0_1

Wiedemann's work is an abridgement. His edition consists of 7,000 words, representing half of the original Arabic text, which is 14,000 words long. Such abridgment is not due to the "verbosity of Arabic" (Sabra 1989, II: lxxxv-lxxxvi). Even if we use the same vocabulary in order to not exaggerate the differences (e.g., "wall" instead of "plane opposite to the plane of the aperture"), it appears that Wiedemann made deep cuts into the text (underlined):

"Eine ähnliche Erscheinung beobachtet man nicht am Mond, weder zur Zeit der Finsternis, noch am Anfang oder Ende des Monates, wenn er sichelförmig ist, trotzdem seine Gestalt der der Sonne im obigen Fall entspricht.

Er erscheint stets rund, auch, wenn die Bedingungen, unter denen die Beobachtung angestellt wird, in beiden Fällen genau gleich sind." (Wiedemann 1914: 156).

"No such thing happens with the eclipse of the Moon, nor in the early or last days of the month when the moon is crescent-shaped, and even though the remaining part of the Sun, when the eclipse is not a total one, resembles the shape of the Moon at the beginning or at the end of the month. Whenever a substantial part of the Sun remains, it looks like a crescent, when the Moon is seen on clear nights. And if, in the early or last days of the month, the Moon is facing a body with an aperture similar to that which produces a crescent-shaped image when the Sun is facing that aperture at the time of its eclipse, and [if] the moonligth appears on the [wall], its light will always be circular. It will never be like the image of the sunlight, even if the two apertures facing the Sun and the Moon are equal."

Although Wiedemann successfully caught the overall content of the work, whole passages were left untranslated. Some other minor problems affect his résumé.

As the manuscripts are not punctuated, his translation does not always keep pace with the Arabic phrasing. Some sentences are cut into two, while clauses of separate sentences are put together. In the following example, the words left untranslated by Wiedemann are underlined (points and commas added):

والسطح المقعر مماس لكُرة القمر على قوس من دائرة محيطة بكُرة القمر. ثم يمتد هذا السطح المخروط المقعر حتّى ينتهي إلى كُرة الشمس. فيقطع كُرة الشمس على قوس من دائرة مساوية للدائرة التي هي قاعدة السطح المحدب، وذلك أن المخروط الذي يحيط بكرة الشمس مساو للمخروط الذي يحيط بكرة القمر، وقد تبين ذلك أصحاب التعاليم. فالسطحان المخروطان المحدب والمقعر قاعدتاهما قوسان من دائرتين متساويتين. (lines 118–25)

"Die konkave Fläche berührt die Mondkugel in einem Kreisbogen, dann schreitet sie bis zur Sonne fort und schneidet sie in einem Kreis, der gleich dem entsprechenden Kreis des konvexen Kegels ist.
Nach den Mathematikern ist der die Sonnenkugel begrenzende Kegel gleich dem die Mondkugel begrenzenden.

Die Sichel ist also von zwei Bögen von zwei gleichen Kreisen begrenzt." (Wiedemann 1914: 157).

"The concave surface is tangent to the Moon sphere, along an arc of circle ... it extends up to the Sun sphere. The Sun sphere is cut off along an arc of a circle, which is the base of the convex surface, because the cones bounding the Sun and the Moon sphere are equal, as was found by the mathematicians. The bases of the two surfaces of the cone, the convex and concave, are two arcs of two equal circles. If we examine the intersection of the convex and concave surfaces of the cone, this makes a crescent-shaped figure bounded by two arcs of two equal circles."

In this passage, Ibn al-Haytham introduces the result of the mathematicians (i.e., the astronomers) to account for the equality of the two [envelope] surfaces of the cones. As the two heavenly bodies have the same angular size, their cones are equal. However, because of the punctuation used by Wiedemann, the causal relationship between the two clauses of this sentence is not rendered (Arabic has وذلك أن...).

In a limited number of cases, we also find accidental misinterpretations of Ibn al-Haytham's intentions. See for example this edition, lines 89–93 (note 24).

The shortcomings that affect this abridged version in no way diminish the value of Wiedemann's work, which has yet to be praised for giving Western scholars access to the content of Ibn al-Haytham's work. I have consulted it on many occasions.

2. Naẓīf's Study

Ibn al-Haytham's *On the Shape of the Eclipse* has been independently studied by Muṣṭafā Naẓīf on the eve of World War II. As an engineer well versed in the history of Arabic science, Naẓīf wrote a penetrating analysis of Ibn al-Haytham's optical research. The work is discussed in the first of his two-volume set (Naẓīf 1942: 182–204; reed. 2008: 276–298).

Which manuscripts did Naẓīf use? His book contains only five passages between quotes (pages 182, 197, 199, 200 and 202), none of which conform exactly to the original. Here is, for instance, a first quotation to the very beginning of the text:

قد يوجد ضوء الشمس وقت كسوفها، اذا نفذ من ثقب ضيق مستدير، وانتهى الى سطح مقابل الثقب هلاليا، اذا كان الجزء الباقى من جرم الشمس هلاليا ولم يستغرق الكسوف جميعها. ولا يوجد مثل ذلك عند خسوف القمر، اذا كان الجزء الباقى منه هلاليا، ولا في أوائل الشهور وأواخرها، بل يوجد ضوؤه أبدا مستديرا اذا كان الثقب مستدير ... الخ (Naẓīf 1942: 182/276)	قد يوجد صورة ضوء الشمس في وقت كسُوفها، إذا خرج ضوءها من ثقب ضيق مستدير وانتهى إلى سطح مقابل الثقب على مثل شكل الهلال، إذا لم يستغرق الكسُوف جميعَها وكان شكل ما بقي منها هلاليًا. وليس يظهر مثل هذه الحال في كسوف القمر ولا في أوائل الشهور وأواخرها إذا كان القمر هلالا وشكل ما تبقى من الشمس. (this edition, Naẓīf's omissions overlined)

The second quotation refers to a passage at the end of the treatise, around time 0.838, just before the beginning of MS P, fol 45v:

ان كل هلال تحيط به قوسان من دائرتين متساويتين فان القوس المقعرة منهما أقل من نصف دائرة، لأن كل دائرتين متساويتين تتقاطعان فان الواصل بين تقاطعهما هو وتر في كل منهما وليس بقطر... (Naẓīf 1942: 197/291)	أن كل هلال يحيط به قوسان من دائرتين متساويتين. فإن القوس المقعرة منهما يكون أقل من نصف دائرة، لأن كل دائرتين متساويتين يتقاطعان، فإن الخط الذي يتصل بين تقاطعهما هو وتر في كل واحد منهما. فهو أصغر من قطر... (this edition)

Although Naẓīf's left-hand excerpts are put in quotation marks, all of them are altered and do not match the handwritten readings. This holds true for all five quotations in his book.

This situation is rather intriguing, given that the right-hand passages agree with each other in all extant manuscripts. There is only one omission in MS L located after the text of the first citation. We also find two minor errors in MSS O and B, before and after the text of the second citation. These readings have no consequences whatsoever on how we are to interpret these passages.

The same applies to diagrams. The diagram that most resembles that of the genuine work is Fig. 10 (Naẓīf 1942: 190). Setting aside the lines of the drawing, only 4 letters out of 14 match that of the Ibn al-Haytham's diagram: these are the letters L N Y T on the vertical axis (Figs. 1.1–1.2).

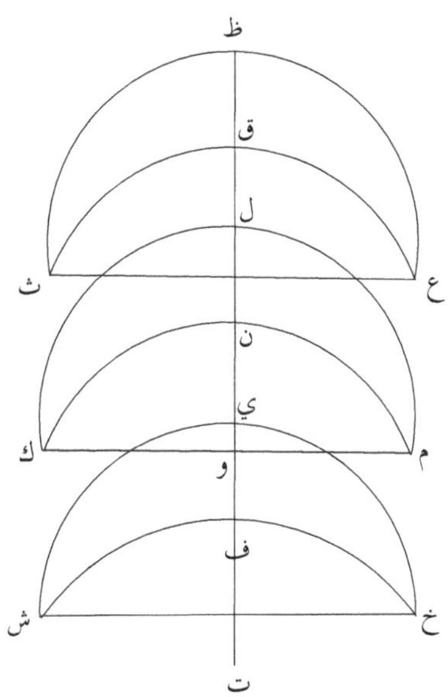

Fig. 1.1. MS P Crescents

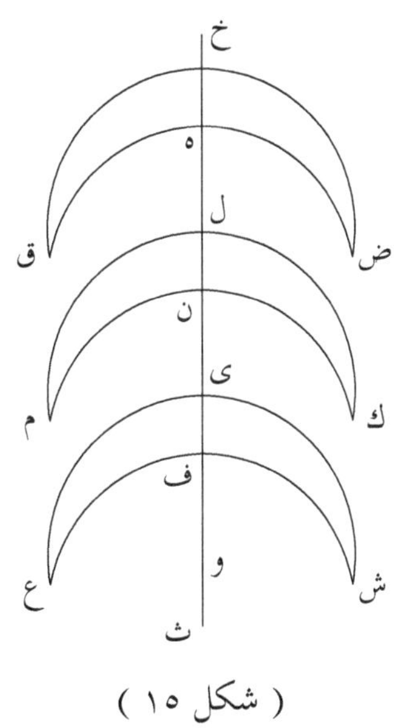

Fig. 1.2. Naẓīf's Crescents

The solution to the puzzle is given at the end of Naẓīf's book. The first thing to note is that Naẓīf had no knowledge of the German translation: he only cites Wiedemann's "Zu Ibn al-Haiṭams Optik" (1910) not "Über der Camera obscura bei Ibn al-

Haiṭam" (1914). Therefore the deviation from Ibn al-Haytham's text is unrelated to that of Wiedemann. Now we note that Naẓīf used Kamāl al-Dīn's *Tanqīḥ al-manāẓir li-dhawi al-abṣār wa al-basā'ir*, a text that perfectly matches his version:

قد يوجد ضوء الشمس وقت كسوفها، اذا نفذ من ثقب ضيق مستدير وانتهى الى سطح مقابل الثقب هلاليا اذا كان الجزء الباقى (١) من جرم الشمس هلالياً ولم يستغرق الكسوف جميعها ولا يوجد مثل ذلك عند خسوف القمر اذا كان الجزء الباقى منه هلالياً ولا في أوائل الشهور وأواخرها (٢) بل يوجد ضوؤه ابدا مستديرا اذا كان الثقب مستدير [٣٨٢] ... الخ (Fārisī 1929: 381–2)

قد يوجد ضوء الشمس وقت كسوفها، اذا نفذ من ثقب ضيق مستدير، وانتهى الى سطح مقابل الثقب هلاليا، اذا كان الجزء الباقى من جرم الشمس هلاليا ولم يستغرق الكسوف جميعها. ولا يوجد مثل ذلك عند خسوف القمر، اذا كان الجزء الباقى منه هلاليا، ولا في أوائل الشهور وأواخرها، بل يوجد ضوؤه أبدا مستديرا اذا كان الثقب مستدير ... الخ (Naẓīf 1942: 182/276)

The only minor differences between the two texts are: 1. the punctuation added by Naẓīf and 2. his removal of the tags introduced by Muṣṭafā Ḥijāzī, when editing Kamāl al-Dīn's text in 1929.

It thus appears that Naẓīf had a second-hand knowledge of *On the Shape of the Eclipse*. He commented on Ibn al-Haytham's work exclusively from the 1929 edition of Kamāl al-Dīn's commentary. This is further confirmed by the list of manuscripts at the end of the book: among the three manuscripts consulted by Naẓīf, none is a witness of *On the Shape of the Eclipse*.[5]

Kamāl al-Dīn al-Fārisī's recension of Ibn al-Haytham's work is not without interest from a scientific point of view. However, the two texts are different: 1. Kamāl al-Dīn produced an original work that went farther than his predecessor's on a number of optical issues; 2. Even when Kamāl al-Dīn cites Ibn al-Haytham, he speaks in his own words: he does not fully and accurately reflect Ibn al-Haytham's thought; 3.

---

5. Istanbul, Ahmet III, MS 3329, fols. 1v–125r: *Commentary on the Almagest* (*Fī Sharḥ al-Majisṭī*); London, India Office, MS 1270, fols. 116v–118r: *On the Compass of Great Circles* (*Maqāla fī birkār al dawāir al-'iẓām*); Lahore, Private Collection 71, fols. 36v–42v: *On Seeing the Stars* (*Maqāla fī ru'ya al-kawākib*). The other two manuscripts appearing in the 2008 reprint are those quoted by Rashed in his introduction: Bursa, Hüseyin Çelebi MS 323, fols. 23v–52r: al-Baghdādī's *On Place* (*Fī al-makān*); Tehran, Majlis-i Shūrā, MS 827: al-Rāzī's *Summary* (*al-Mulakhkhaṣ*).

Muṣṭafā Naẓīf used the Hyderabad edition at a time when the autograph of Kamāl al-Dīn's *Tanqīḥ* was unknown (this is now the Adilnor MS, Malmö). We now know that the Hyderabad edition is not faithful to the autograph manuscript, which has been discovered in the meantime (Sabra 1989, II: lxxii).

Despite this substitution—most likely determined by the troubled times in which this research was done—Naẓīf perfectly rendered the optical and mathematical ideas embedded in this work (1942: 196–8).

3. Sabra's Comment

Another comment on Ibn al-Haytham's *On the Shape of the Eclipse* appeared in the introduction to the English translation of Ibn al-Haytham's *Optics*, Books I-III. When reviewing Ibn al-Haytham's optical works, Abdelhamid I. Sabra devoted three pages to this work (Sabra 1989, I: xlix-li). He gave a good summary of the work with helpful notes on several concepts, such as the word image, to be found in Ibn al-Haytham's treatise. Sabra also used a wider range of sources. According to his own words, he consulted three manuscripts: Istanbul, Fātiḥ 3439, Leningrad B 1030 and India Office 1270, along with commentaries, namely those by Kamāl al-Dīn (1929, II: 381–401), Wiedemann (1914) and Naẓīf (1942). Though short, Sabra's commentary is reliable and well informed. Several problems remain nonetheless:

1. A comment is not a critical edition of the text;

2. All of the manuscripts were not consulted. In particular, Oxford, Bodleian, MS Arch. Seld. A32 (which is the only manuscript that makes extensive use of diacritical marks) and London, India Office, MS 461 (which is one of the rare manuscripts to have the diagrams carefully drawn) were not available to him;

3. Sabra did not inform us how he decided between the different versions he used, that is, firstly, between the three manuscripts he consulted, and secondly, between those manuscripts and the scholarly literature. He simply warns us that:

> Throughout his commentary, Kamāl al-Dīn distinguished the statements which he derived from the *Optics* by introducing them with 'he said', while introducing his own comment with 'I say'. This has sometimes given the impression that he was quoting I.H.'s actual words when in fact he was summarizing or re-phrasing the text (Sabra 1989, II: lxxii).

The same holds for *On the Shape of the Eclipse.* So Sabra probably diverged from Kamāl al-Dīn's version and Naẓīf's commentary on a number of issues, but we do not know exactly on what points he differed.

This survey reaches a simple conclusion: there is so far no reliable edition of Ibn al-Haytham's *On the Shape of the Eclipse.* Thus a critical edition is needed.

2. Codicological and Stemmatological Notes

1. Authorship and Date

Little is known about Ibn a-Haytham's life. He was born in Baṣra in the mid-tenth century—note that the date A.H. 354/965, frequently reported in the literature, refers to Muḥammad ibn al-Ḥasan ibn al-Haytham's birth. Ibn a-Haytham went to Egypt with a plan to control the flow of the Nile that he proposed to the caliph al-Ḥākim.[6] Ibn al-Haytham realized his project was unworkable and admitted failure. According to certain bio-bibliographers, as he feared revenge of the caliph, he preferred to retire by feigning madness. After the ruler's death, most sources agree that he settled next to the Azhar mosque-university in Cairo, earning his living from copying mathematical texts. Ibn al-Haytham died on or after A.H. 432 (11 September 1040–30 August 1041) for Ibn al-Qifṭī stated that "he had in his possession a volum (*juz'*) of geometry written by Ibn al-Haytham's hand and dated A.H. 432" (ed. Lippert 1903: 167; Sabra 1998; Rashed 1993). Historical sources are inconsistent on many other facets of his life.

As is well known, there has been a twenty-year long controversy between Abdelhamid I. Sabra and Roshdi Rashed on whether we should or should not identify Abū 'Alī ibn al-Ḥasan ibn al-Ḥasan ibn al-Haytham (the mathematician) with Muḥammad ibn al-Ḥasan ibn al-Haytham (the philosopher). This controversy casts a shadow over some major episodes of Ibn al-Haytham's life and works. The two scholars have

6. Abū 'Alī Abū Manṣūr Tāriq al-Ḥākim (called al-Ḥākim bi Amr Allāh) reigned from 29 Ramaḍān A.H. 386/15 October 996 to 27 Shawwāl A.H. 411/13 February 1021 (Canard 1975: 79–84). Ibn al-Qifṭī constantly refers to him as "ruler" (*al-ḥākim*) without explicitly mentioning his identity, which is inferred only on the basis of his "cruelty and versatility."

used different methods: Sabra supports the identity of the two persons on the basis of biographical interpolation (Sabra 1972: 189–96; 1989, II: xix–xxxii; 1998: 1–50; 2002/3: 95–108); Rashed supports his view on the basis of internal analysis of Ibn al-Haytham's works (Rashed 1993: 1–19, 490–491, 511–538; 2000, 937–941; 2002: 957–959; 2007: 47–63). Altogether, over 160 pages are devoted to the debate. This knotty controversy looks like a dilemma: either we know with certainty only a few facts about Ibn al-Haytham's life and works, or we know a larger amount of facts which are uncertain in nature. Neither position is comfortable according to the standards of history of science.

Because of this controversy, the authorship of *On the Shape of the Eclipse* needs to be established. A first response is given by the way all manuscripts are titled: "al-Ḥasan ibn al-Ḥasan ibn al-Haytham's *Epistle on the Shape of the Eclipse*." The text provides additional information. In the course of the treatise, the author says that he has dealt with the rectilinear propagation of light in his *Optics*: "From every point of every self-luminous body, light radiates in every straight line ... We have explained that, with due proof and experimentation, in the first book of our work on *Optics*" (lines 47–8). We also read in the subsequent text: "This has been shown in the first chapter of the book of *Conics*" (line 56), a mention that fits well within the focus of interest of Ibn al-Haytham, who attempted a reconstruction of Apollonius' *Conics* lost Book VIII (for editions of this work, see Hogendijk 1985; Rashed 2000).

These facts indicate with no doubt whatsoever that the author of *On the Shape of the Eclipse* was Abū ʿAlī al-Ḥasan ibn al-Ḥasan ibn al-Haytham, the mathematician and astronomer.

The date of composition of *On the Shape of the Eclipse* is unavailable using common historical methods. None of the extant five manuscripts bears a date of composition and, to my knowledge, there is no historical document usable for dating. However, it appears that the date of composition of this work can be determined through astronomical methods. As the demonstration is too long to fit into the main text, I will content myself with referring the reader to the *Appendix: A Tentative Astronomical Dating for the Epistle*. The result of this research is that Ibn al-Haytham may have recorded in his treatise the partial solar eclipse visible on 28 Rajab A.H. 380/21

October 990 CE from Baṣra. Should this dating be confirmed by future research, *On the Shape of the Eclipse* would definitely be an early work by Ibn al-Haytham—a fact that fits well with the case that the author of the work is "al-Ḥasan ibn al-Ḥasan ibn al-Haytham," with no mention of his future *kunya* Abū ʻAlī.

Since the *Optics*, Book I, is referred to in *On the Shape of the Eclipse*, the above dating would mean that Ibn al-Haytham started composing *Optics* in his youth, before leaving the Būyid ʻIrāq around the turn of the millenium. Thus his optical research presumably took shape under the reign of the emir Bahā' al-Dawla (wa-Diyā' al-Milla), who ruled over ʻIrāq from 989 to 1012 after his brother Ṣamṣām al-Dawla (Shams al-Milla), who reigned from 983 to 986 (for more details on the Būyid dynasty, see Bosworth 1975; Kraemer 1992; Donohue 2003).

2. The Manuscripts

Ibn al-Haytham's *On the Shape of the Eclipse* is extant in five manuscripts:

F Istanbul, Süleymaniye, Fātiḥ, MS 3439, fols. 117r–123v. Size 190 × 130 mm. Incomplete. The manuscript, written in poor *naskh*, was completed by Ibrāhīm ar-Rūjānī al-Bakrī in Mosul, in the night of ʻAshūrā' AH. 587/7 February 1191 (Krause 1936: 458). The text is unfaithful up to time 0.061 and ends at time 0.922. The diagrams are legible enough, but often distorted. Diagram 4 is missing.

B Oxford, Bodleian Library, MS Arch. Seld. A.32, fols. 81v–100v. Size 180 × 115 mm. The copy, written in a careless *naskh*, "was transcribed before A.H. 633 (A.D. 1235–6), being contained in a volume which came into the possession of Yaḥyā ibn Muḥammad ibn al-Labūdī [of Damascus] in that year. In the colophon we are informed that the copyist transcribed the text from a copy claiming to have been transcribed from the prototype" (Sabra and Shehaby 1971, ix). The diagrams often impinge on the text. They are distorted by a lack of parallelism and squareness. The intersection points are rough.

P St. Petersburg, Institute of Oriental Manuscripts, MS B 1030, fols. 21r–47v. Size 170 × 92 mm. MS P predates the mid-fourteenth century. We know that this manuscript was copied and checked against Ibn al-Haytham's autograph in A.H. 750/1349. "This collection, written in mediocre *nastaʻlīq*, is of great scientific quality" (Rashed 2005: 15). The text ends at time 0.926. All of the diagrams are geometrically clear and accurate.

O London, India Office, MS 1270 (Loth 734), fols. 79r–86v.[7] Size 279 × 114 mm. The manuscript is in good *naskh*, evidently from the sixteenth century. As stated by Loth, this copy is "well written in a small hand, with numerous neatly drawn diagrams" (Loth 1877: 214). MS O was initially part of the library of Richard Johnson (1753–1807), who came back to England in 1799. MS O was purchased by the India Office at the nabob's death in 1807. The text ends at time 0.998, seven words before the end.

L London, India Office, MS 461 (Loth 767), fols. 8v–34r. Size 229 × 140 mm. The manuscript is written in good *nastaʿlīq*. The date of the manuscript is deducible from the fact that the copy of al-Ṭūsī's *Treatise on Astrolabe* (*Risāla al-asṭurlābiyya*), appearing on fols. 1–7, was revised on 14th Shawwāl A.H. 1198/31 August 1784. The manuscript could have belonged to Governor-General Warren Hastings (1773–85), before it passed to the London Library (Loth 1877: 223). This manuscript has finely drawn diagrams, which all appear on a separate sheet.

Once acquired, the copies of the five extant manuscripts of *On the Shape of the Eclipse* were collated.

A first methodological novelty was to cut the digital copies into strips to lay the entire text on a single line. The five manuscripts were paralleled through a drawing software, whose baseline was calibrated by attributing the value 0.000 to the beginning of the text and the value 1.000 to its end. This turned out to be an effective device to compare the manuscripts and track the different readings, because any deviation in any manuscript can be identified easily and its point of occurrence in the edited work designated. This device facilitated the collation of the text and the tracking of handwritten variants.

The second methodological innovation was to apply the same editing rules to the text and diagrams—a suggestion that has been recently brought forward in view of the discrepancies that are often seen between the diagrams found in manuscripts and those depicted in scholarly editions of the same works: "The diagrams should be presented as they are found in the MSS, accompanied by a critical apparatus ... Where this is possible, we should seek to establish the text history of the diagrams and

7. O for Oblongus.

present this in a stemma" (Sidoli 2007: 546). In view of the advances made in diagram studies[8] and digital stemmatology,[9] a new method for the critical editing of geometric diagrams was devised—a method which, to my knowledge, has never been applied. As the stemma codicum of *On the Shape of the Eclipse* has been established elsewhere, I will limit myself to summing up the main results, while referring the reader to this publication for details (Raynaud 2014*c*).

3. The Stemma of the Text

As is well known to philologists, long omissions are of special interest for building the stemma codicum (Viré 1986; Woerther and Khonsari 2001). While a scribe can compensate for the omission of one word, he cannot restore a long passage without referring to the source. Thus all the descendants of a corrupted model will carry the same corruption. These omissions were detected across the manuscripts.

Then, I encoded all text accidents in a matrix of characters, which consists of six taxa (the five MSS and the out-group, that is, the text without errors) and of as many characters as there are omissions in the text from time 0.000 to time 0.922, a date that corresponds to the end of MS Fātiḥ. After comparing and examining the various programs available, I decided to use PHYLIP 3.69 (Felsenstein 2009), a program providing a whole package of algorithms.

The major cladistics techniques have recently been compared by estimating the similarity between the stemmata they provided on three independent data sets (Roos and Heikkilä 2009). This comparison showed the advantage of the Maximum Parsimony and RHM methods. In view of its wider diffusion, I have used the first one.

8. See Cambiano (1992), Decorps-Foulquier (1999), Netz (1999), De Young (2005), Mascellani et al. (2005), Crozet (2005), Saito (2005), Saito (2006), Sidoli (2007), Manders (2008), Sidoli and Saito (2009), Jardine and Jardine (2010), Saito et al. (2011), Sidoli and Li (2011), Saito and Sidoli (2012), Mumma et al. (2013).

9. The cladistic analysis is now being used in the critical editing of both literary and scientific texts. Key studies are: on the approach in general (Glenisson 1979; Reenen et al. 1996; Robinson 1996; Dees 1998; Huygens 2001; Woerther and Khonsari 2001; Macé et al. 2001; Reenen et al. 2004; Macé and Baret 2006; Cipolla et al. 2012); in literary texts (Robinson et al. 1996; Salemans 1996; Barbrook et al. 1998; Salemans 2000; Mooney et al. 2001; Windraw et al. 2008; Maas 2010); and in scientific texts (Brey 2009; Pietquin 2010; Cardelle de Hartmann et al. 2013).

When PHYLIP's Maximum Parsimony Algorithm is applied to the matrix of characters, a single most parsimonious tree is found (Fig. 1.3). The manuscripts connect to several ancestors by means of branches whose (horizontal) length is proportional to the number of transformed characters between the ancestor and the manuscript. A branch of zero length means that there is no difference between the manuscript (terminal node) and the progenitor (intermediary node).

```
+outG
|
1----Petersbg
|
|                     +--India1270
+---------------------2
                      |  +------------Bodleian
                      +--3
                         |         +------------------Fatih
                         +---------4
                                   +---------------------India461
```

Fig. 1.3. The Text Stemma

In this stemma codicum, MS Petersburg is directly connected to the ancestor [1]. MSS India Office 1270 and Bodleian connect to intermediate nodes [2] and [3]. MSS Fātiḥ and India Office 461 have a common intermediate ancestor [4]. The stemma helps us to decide which lectio should be followed when the manuscripts disagree. This recommends using preferably MSS Petersburg (the *lectio praeferenda*) and India Office 1270, for critically editing the text of *On the Shape of the Eclipse.*

4. The Stemma of Diagrams

The analysis of the handwritten diagrams follows in the same footsteps. The diagrammatic errors can be classified (a line drawn/missing; a geometric property true/false; a diagram oriented clockwise/counter-clockwise, etc.), and detected throughout the manuscripts by visual inspection.

Then a matrix of characters is made of six taxa (the five extant manuscripts of Ibn al-Haytham's work and the out-group, which simply consists of the list of common features) and of the 96 errors that we have detected in diagrams.

When PHYLIP's Maximum Parsimony Algorithm is applied to the matrix of characters recording the errors defined and surveyed, again one most parsimonious tree is found (Fig. 1.4).

```
+outG
|
1-----Petersbg
|
|              +----------India461
+--------------2
               |         +------India1270
               +---------3
                         |            +----------------Bodleian
                         +------------4
                                      +-----------------Fatih
```

Fig. 1.4. The Diagram Stemma

In the stemma codicum above, the closest fit is MS Petersburg, which is again very similar to the out-group. MS Petersburg directly connects to the common ancestor [1]. MS India 461 stems from an intermediate node [2] between MSS Petersburg and India Office 1270. MSS Bodleian and Fātiḥ share a common intermediate ancestor [4]. This recommends using preferably MSS Petersburg, then India Office 461 and India Office 1270, for critically editing the diagrams, while MSS Fātiḥ and Bodleian are inappropriate in this respect.

The main differences between the text and diagram stemmata arise because the diagrams are embedded in the text in all manuscripts, except in MS L, where the scribe redrew them on a separate sheet—conceivably with some critical goal in mind. Once MS L is removed from the data, the two stemmata are virtually identical.

3. Editorial Procedures

1. Scientific Vocabulary

Ibn al-Haytham's phrasing is fully explicit, a trait that may appear redundant to more than one reader. Faced with this situation, Wiedemann chose to shorten Ibn al-Haytham's text by encoding repeating words with symbols, such as *S* (*Sonne*), *L*

(*Loch*), *W* (*Wand*) and so on. I have opted for a literal translation instead because, given that scholars disagree about Ibn al-Haytham's legacy, one purpose of this edition is to prepare a comparison of the texts available on the camera obscura. I have reworked the text—mainly syntaxically—only when the *ad verbum* translation failed to render the meaning of the text. Any deviation from the original is pointed out in the critical apparatus on each occasion.

As is well known to translators, any piece of text can be rendered in different language registers. The common language register is appropriate for a translation made *de verbo ad verbum*. However, this register is often out of step with optical and scientific vocabulary and causes difficulties in reading. The formal language register is needlessly sophisticated and leads to anachronisms such as suggesting that Ibn al-Haytham was aware of optical concepts that were discovered much later. These considerations suggest excluding these two solutions. I have translated *On the Shape of the Eclipse* keeping in mind an intermediate language register, which is neither too common, nor too formal. This solution allows for a global paralleling of the Arabic and English texts, while an editor's task focus on disambiguation. When there is no possible confusion, my translation is literal. For example نصف قطر (*niṣf quṭr*) is translated as "semi-diameter" rather than "radius" (e.g., line 180). When potential misunderstanding exists, I have split the term into different words.

Mathematics. The word سهم (*sahm*), literally "arrow," occurs frequently in Ibn al-Haytham's text. The term describes the "versine" when applied to the distance from the chord to the arc of a unit circle, the "sagitta" when the circle is not a unit circle, e.g., line FR of arc ŠFẖ (lines 623). The same word can refer to the "axis" ṢṬF of the light cone spread from the aperture (lines 130, 139) and, more critically, it can also refer to the "generatrix" SḤT of the same cone (line 135). Here any word-to-word translation would fail to make the text clear. In view of this, the word *sahm* has been translated to "sagitta," "axis" or "generatrix," according to the context.

Optics. The adjective الهلال المضيء (*al-hilāl al-muḍī'*) has many occurrences in Ibn al-Haytham's text. It has been split into two for the sake of disambiguation. I have translated it as "self-luminous crescent" (line 60) when the term unequivocally refers to the remaining part of the Sun during the eclipse, while preferring "light crescent" (line 358) when the term refers to the image cast on the projection plane of the darkroom.

The terms have been translated consistently, so that the reader can track a word in the text from the Arabic-English Glossary appended to this book.

2. Spelling Variants

All five manuscripts have spelling variants. We can often find ضؤ or ضو instead of ضوء (*ḍaw'*, "light"); علي instead of على (*'alā*, "to"); اوايل instead of أوائل (*awā'il*, "beginning"). The verbal forms تكون، يكون (*yakūn/takawwūn*, "he/she is") and تخرج، يخرج (*yakhruj/takharruj*, "he/she comes out") are muddled throughout the manuscripts. Affixe pronouns are juxtaposed rather than attached: we read رأس هما instead of رأسهما (*rāsahumā*, "their two apices"). All those peculiarities have been erased, while other spelling variants have been kept as long as the understanding of the text and ease of reading were not at stake.

3. Punctuation

None of the manuscripts is puncutated with the use of commas, semicolons and full stops. Punctuation is basically limited to the division into sections, for which the scribes have used the signs ∴ and ه (short and long pauses). At the risk of losing a little freedom in the rendering of the text, I have proceeded codicologically: a line break was used to introduce a new paragraph when at least three out of five manuscripts were in accordance in the identification of a break.

Within each paragraph, the Arabic is just articulated by the use of the conjunction (in general verb+و), causal particle (verb+ف), expression of consecution (ثم) and pronouns introducing the subordinate clause. For the ease of reading and correspondence with the English translation, the Arabic text has been cut into clauses and propositions.

4. Diacritical Marks

The manuscripts have only a few diacritical marks. Unfortunately, this lack particularly affects MS P, which is virtually free of any diacritics. In MS L, diacritics are mostly limited to the use of the *shadda* ّ to represent the gemination of consonants. We find in Ibn al-Haytham's text several occurrences of بعد. The different meanings of this skeleton are distinguished by the diacritics alone: after (بَعْدَ), still, yet (بَعْدُ), to be away (بَعُدَ), remote (بَعِيد), distance (بُعْد). MS B provides the forms بَعد (after), بَعُد (to be or move away), بُعد (distance), when the other manuscripts generally lack diacritics. I have used the diacritical marks sparingly to solve an ambiguity.

4. Tips on Reading

This book is made up of two parts: the Critical Edition of the Arabic text with English Translation (Chapter 2) and the Commentary (Chapters 3–4).

The Critical Edition uses two types of references: the minor clarifications and bibliographic references appear in footnotes; all significant issues that call for a substantial development are dealt with in the Commentary.

A cross-referencing system has been put in place between the Critical Edition and the Commentary. In Chapter 2, at the start of an important passage, a footnote indicates the page of the Commentary where the passage is commented on. Conversely, in Chapters 3–4, any discussion is referred to the lines of the Critical Edition.

1. General Outline of the Treatise

Chapter 4.3 (pp. 116–30) provides the reader with a general outline of the demonstration to easily follow Ibn al-Haytham's reasoning. In *On the Shape of the Eclipse*, he achieves a number of notable optical results, through the handling of simple mathematical relationships, such as ratios and proportions in similar and opposite triangles. This enables him to study experimentally the camera obscura.

1. Ibn al-Haytham demonstrates the *inversion of the image* in the darkroom. This property was known to al-Kindī and earlier Chinese scholars—for example, the *Mo Ching* states: "The image is inverted (到 *dao*) because of the intersection (午 *wu*). The intersecting place is a point (端 *duan*)" (Needham 1962: 82). The only substantial difference with al-Kindī and the Chinese is that Ibn al-Haytham approaches *geometrically* the image inversion (Chapter 4.2, pp. 114–6).

2. He investigates the shape of the image *as a function of the size of the aperture*: the wider the aperture, the rounder the image. He goes even further and notices that the concave side of the image is not of perfect circularity: the solar image is flattened, and this flattening depends on the magnitude of the eclipse and radius of the aperture (Chapter 4.4, pp. 130–42).

3. He studies the shape of the image *as a function of the focal distance* of the camera obscura: the wider the focal distance, the wider the crescent-shaped image (Chapter 4.5, pp. 142–5).

4. He studies the image *as a function of the shape of the aperture* and shows that it has influence only if the aperture is large. When equipped with a pinhole, the dark chamber produces a sharp picture: there is correspondence between image points and object points. When the aperture is enlarged, the image cast on the screen can be decomposed in as many overlapping patches of light as we want, each of them being the projection of the solar image through one point of the aperture. Whatever the shape considered, there is a rounding of the image (Chapter 4.6, pp. 145–7).

5. Finally, he devotes a long passage to the study of the shape of the image *as a function of the light source*, by comparing how the images of the Sun and the Moon appear in the darkroom. He puts forward two arguments, one of them is mistaken and the other correct. He mistakenly claims that the ratio required for light to appear crescent-shaped is not met in the case of the Moon. However, he appropriately explains that the image of the Moon is even more rounded than that of the Sun, because its faintness leads to the vanishing of the tips and edges that receive less light than the center of the image. In this respect, it is safe to say that Ibn al-Haytham took a first step towards proto-photometry (Chapter 4.7, pp. 148–59).

2. Reading the Diagrams

The diagrams of *On the Shape of the Eclipse* have been critically edited. The diagrams thus retain the original proportions found in manuscripts. Since the overall device cannot be depicted to scale—the Sun-Earth distance is about 10^{10} times more than the focal distance of the darkroom—there is no perfect picture. The reader is here provided with a three-dimensional view of the device. (For convenience, the bodies' positions are invariant throughout the book: the celestial bodies are on the left; the aperture in the middle; the images cast on the right.)

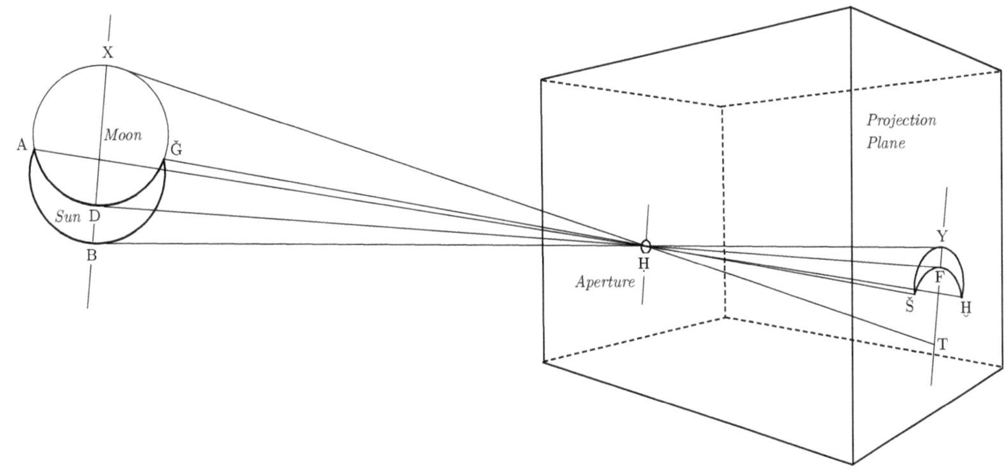

Fig. 1.5. Overview of Ibn al-Haytham's Device

During the partial solar eclipse, the Sun ABǦ is covered by the Moon ADǦ. What remains of the Sun is the crescent-shaped figure ABǦD. When the aperture reduces to a pinhole (Fig. 1.5), the crescent-shaped figure ABǦD casts through Ḥ the image ŠYḪF on the projection plane. This is the result of central symmetry.

When the aperture is large (Fig. 1.6), the crescent-shaped figure ABǦD casts through the aperture HṬḤ a composite image, which is made by the overlapping of image ŠYḪF obtained through Ḥ, image KLMN obtained through Ṭ, image ṮẒGQ obtained through H, and as many images as one wishes, by taking any other point of the aperture. The resulting image of the Sun is the aggregation of all these elemental images.

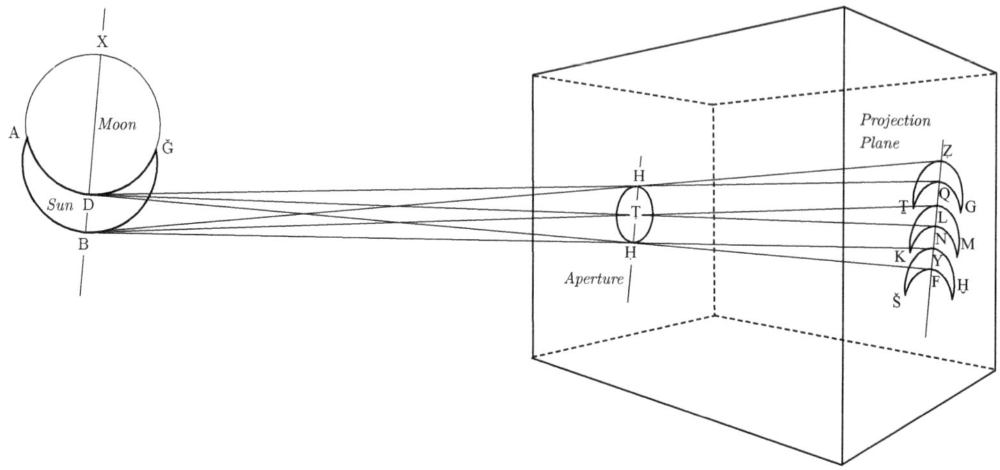

Fig. 1.6. The Formation of the Image

Diagram 1 (Plates, p. 261) represents the three crescent-shaped figures one above the other as in Fig. 1.6. Diagram 2 is a lemma stated in view of the next proposition. Diagram 3 reproduces Diagram 1 with some additional features showing how the crescent-shaped figures overlap one another. Diagram 4 is a detailed view of the two crescent-shaped figures ŠYḪF and KLMN in Diagram 3.

5. Transliteration

Given the large number of points in Ibn al-Haytham's diagrams, I have decided not to transliterate the Arabic letters according to the Kennedy-Hermelink system (Kennedy 1991–2), because beyond the 20th Arabic letter, arbitrary Latin and Greek letters need to be introduced. I have adopted DIN-31635 instead, except for the letter G ع, which does not interfere with Ğ ج and Ġ غ.

Z ز	W و	H ه	D د	Ğ ج	B ب	A ا
N ن	M م	L ل	K ك	Y ي	Ṭ ط	Ḥ ح
Š ش	R ر	Q ق	Ṣ ص	F ف	G ع	S س
Ġ غ	Ẓ ظ	Ḍ ض	D̲ ذ	Ḫ خ	T̲ ث	T ت

5. Sigla

The following abbreviations are used:

add.	addidit
ante	ante
corr.	correxit
del.	delevit
marg.	in margine
lac.	lacuna
om.	omissit
post	post
rep.	repetivit
scr.	scripsit
s.p.	sine punctis
tr.	transposuit
?	lectio incerta
!	sic
<>	addendum
[ ]	omittendum
$^{(}\ldots^{)}$	parenthesis added around a manuscript text to identify ratio terms
$\ulcorner^{\text{om.X}}$	beginning of a long omission in MS X
$\urcorner^{\text{om.X}}$	end of a long omission in MS X
$\vert^{\text{X}}$	short pause marked by the sign ∴ in MS X
$\Vert^{\text{X}}$	long pause marked by the sign ه in MS X

Chapter 2
Arabic Text and Translation

D. Raynaud, *A Critical Edition of Ibn al-Haytham's On the Shape of the Eclipse*,
Sources and Studies in the History of Mathematics and Physical Sciences,
DOI 10.1007/978-3-319-47991-0_2

F MS Fātiḥ
B MS Bodleian
P MS Petersburg
O MS India Office 1270
L MS India Office 461

F117r O79r P21v B81v L8v

In the name of God, the Merciful[1]

al-Ḥasan ibn al-Ḥasan ibn al-Haytham's Epistle on the Shape of the Eclipse

<1. The Observations>

<1.1 Effect of the Size of the Aperture on the Image of the Sun>

A crescent-shaped image of the light of the Sun can be seen[2] at the time of the eclipse, if its light has passed through a narrow circular aperture and has reached[3] a plane opposite the aperture, provided the eclipse is not a total one and the shape of its remaining part <of the Sun> is crescent-shaped. No such thing happens with the eclipse of the Moon, nor in the early or last days of the month[4] when the Moon is crescent-shaped, and even though the remaining part of the Sun, when the eclipse is not a total one, resembles the shape of the Moon at the beginning or at the end of the month. Whenever a substantial part of the Sun remains, it looks like a crescent <Moon>, [F117v] when it is seen on clear nights. And if, in the early or last days of the month, the Moon is facing a body with an aperture similar to that which produces a crescent-shaped image when the Sun is facing that aperture at the time of its eclipse,

---

1. O add. "Glory to Allah" and B add. "Help me Lord."
2. Lit.: "There may be a crescent-shaped image..."
3. Lit.: "has ended up to..."
4. Lit.: "months" in plural, that is, "when the Moon is waxing or waning."

F117r O79r P21v B81v L8v

بسم اللّٰه الرحّمن الرحّيم

مقالة للحسن بن الحسن بن الهيثم في صورة الكسوف

قد يوجد صورة ضوء الشمس في وقت كسُوفها، إذا خرجَ ضوءها من ثقب ضيق مستدير وانتهى إلى سطح مقابل الثقب على مثل شكل الهلال، إذا لم يستغرق الكسُوف جميعَها وكان شكل ما تبقى منها هلاليًا. وليس يظهر مثل هذه الحال في كسوف القمر ولا في أوائل الشهور وأواخرها إذا كان القمر هلالا، وشكل ما تبقى من الشمس، إذا لم يستغرق الكسُوف جميعها، يشبه شكل القمر في أوائل الشهور وأواخرها. وإذا ⌐omL كان ما تبقى من الشمس جزءًا مقتدرًا، فهو يشبه الهلال، إذا مضت له F117v ليالي يسيرة. وإذا ¬omL قوبل القمر في أوائل الشهور وأواخرها بجسم فيه ثقب مثل الثقب الذي يظهر منه صورة الهلال عند مقابلة الشمس بذلك الثقب في وقت كسوفها، وظهر ضوء القمر على سطح مقابل الثقب انّما يوجد الضوء

---

1 post الرحّيم] العزّة للّٰه add O رب أعن add B || 2 في] om L || 6 اذا كان القمر ... واواخرها] marg P || 8 وإذا كان ما تبقى ... ليالى يسيرة. وإذا] 16 words om L by homoioarkhton

and <if> the moonlight appears on the plane opposite the aperture, its light will always be circular. It <the moonlight> will never be like the image [L9r] of the sunlight, even if the two apertures facing the Sun and the Moon are equal, [P22r] and if the distance that separates the two planes displaying the light from the two planes of the apertures[5] are equal on both occasions. |[PO]

And this is true[6] in the event the eclipse of the Moon is not a total one and what remains of it is crescent-shaped; its light coming from the aperture is not crescent-shaped. |[O]||[B]

However, [B82r] the image that results from the light of the Sun at the time of its eclipse only appears such <i.e. crescent-shaped> if the aperture is narrow and up to a certain extent of capacity. Then, if the aperture is expanded further,[7] the image of the shape appearing on the plane opposite the aperture is changed. And the more increased the capacity of the aperture, the more changed the image. This change continues until some limit of capacity of the aperture, at which the figure that appears in the light emerging from the aperture is no longer crescent-shaped, but circular.[8] Subsequently, no matter how large the aperture afterwards, the light will always be displayed as circular provided that the aperture is circular. |[O]||[B]

---

5. Lit.: "the two planes of the two apertures."

6. Lit.: "Similarly in the case of the Moon, at the time of the eclipse..."

7. This notion already appears in Pseudo-Aristotle's *Problemata Physica* XV, 11, where one reads: "διὰ μεγάλων ὀπῶν," "per foramina ampliora," and also, with a little more doubt, "<μικρ>αὶ ... ὀπαὶ γίνονται," "<parva> ... foramina fiunt."

8. Lit.: "and the light that comes out from the aperture becomes circular."

أبدًا مستديرًا. ولا يوجد في وقت من الأوقات على مثل الشكل الذي يظهر من ضوء L9r الشمس، وان كان الثقبان اللذان يقابلان بهما الشمس والقمر متساويين، وكان بُعد السطحين P22r اللذين يظهر عليهما الضوء عن سطحي الثقبين في الوقتين متساويين. |PO

وكذلك حال القمر في وقت كسوفه، إذا لم يستغرق الكسوف جميعه وكان ما يبقى منه على مثل شكل الهلال، ليس يكون ضوءه الذي يخرج من الثقب هلاليًا. |O||B|

ومع ذلك B82r فإن الصورة التي تظهر من ضوء الشمس في وقت كسوفها إنما تظهر إذا كان الثقب ضيقًا وإلى حد من السعَة. ثم إذا وسع الثقب، تغيرت صورة الشكل التي يظهر على السطح المقابل الثقب. وإذا زيد في سعة الثقب زاد التغير الذي في الصورة. ثم يتمادى هذا التغير إلى أن ينتهي الثقب إلى حد من السعة، فتبطل الهلالية التي يظهر في الضوء ⌐omL الذي يخرج من الثقب ويصير ⌐omF الضوء الذي يخرج من الثقب ¬omFL مستديرًا. ثم كلما وسع الثقب ¬omF بَعدَ ذلك ظهرَ الضوء أبدا ⌐omF مستديرًا ¬omL إذا كان الثقب مستديرًا. ⌐omL |O||B|

14 يقابلان] يقابلان FO يقابل BPL || 15 عليهما] يظهرما F || 15 سطحي] سطح O || 20 تظهر] يظهر BOL the verb تظهر follows the fem sing الصورة || 21 تظهر] يظهر BOL || 22 صورة] صورت B || 24 الذي يخرج من الثقب ويصير الضوء الذي يخرج من الثقب] 10 words om L || 25 الضوء الذي يخرج من الثقب] 5 words om F || 26 بَعدَ ذلك ظهرَ الضوء أبدا] 5 words om F || 26 إذا كان الثقب مستديرًا] 4 words om L

If one experiments[9] with sunbeams[10] emerging from wide apertures at the time of the solar eclipse, they will always appear similar to the shape of the apertures: If the wide aperture is circular, the light that comes out from it will appear circular; If the aperture is square,[11] the light [P22v] coming out from it will appear square, [L9v] and so on, with any other shape of the aperture, provided that it is wide. The shape of the sunlight that comes from it at the time of the eclipse will be as the shape of the aperture, provided that the plane displaying[12] the light is parallel to the plane of the aperture. |[P]||[BO]

<1.2 A Different Observations in the Case of the Moon>

But if the moonlight[13] comes out from one of these apertures and appears on the plane parallel to the aperture, and if the aperture is circular, the light will only be circular,[14] whenever the aperture is narrow, the Moon is full or crescent-shaped, as is the case within the first or last days of the month, or at the time of an eclipse. Likewise, while the moonlight comes out from apertures of different forms, the image of its shapes will be similar to the shapes of the apertures. |[P]||[BO]

---

9. On the verb to experiment (*i'tabara*) see Commentary, pp. 95–6.

10. Lit.: "If the lights of the Sun emerging... are experimented." Light in plural denotes the rays of the Sun observed in different situations. See Commentary, p. 145.

11. This passage is reminiscent of Pseudo-Aristotle's *Problemata Physica* XV, 6: "Why is it that when the Sun passes through quadrilaterals, as for instance in wickerwork, it does not produce a figure rectangular in shape but circular?" (trad. Hett 1936: 333). This notion again appears in Book XV, 11 (1936: 341). The Greek and Latin wordings are: "διὰ τί δ' ἥλιος διὰ τῶν τετραπλεύρων διέχων οὐκ εὐθύγραμμα ποιεῖ τὰ σχήματα, ἀλλὰ κύκλους," "Cur sol, per quadrangula penetrans, non figuras rectis lineis terminatas, sed circulos efficit?" The notion also appears in Book XV, 11: "δι' οπῆς ἐὰν λάμπῃ εὐγονίου τὸ φῶς," or "per angulatum foramen lux illuscescit."

12. Lit.: "on which the light is displayed..."

13. Lit.: "As for the moonlight, if it comes..."

14. The idea of comparing the Sun and the Moon also goes back to *Problemata Physica* XV, 11: "ἀπὸ δὲ τῆς σελήνης οὐ γίνονται, οὔτ' ἐκλειπούσης, οὔτ' ἐν αὐξήσει οὔσης, ἢ φθίσει," "Luna autem huiusmodi lunulas non producit, neque deficiens, neque increscens, neque decrescens."

وإذا اعتبرت أضواء الشمس التي تخرج من الثقوب الواسعة في وقت كسوف الشمس، وجدت أبدا على مثل أشكال الثقوب: إن كان الثقب الواسع مستديرًا ⌐omF ظهر الضوء الذي يخرج منه ¬omF مستديرًا، وإن كان ⌐omF الثقب مربعًا ظهر الضوء الذي P22v يخرج منه مربعًا، L9v وبأي شكل كان الثقب، إذا كان واسعًا. فإن شكل ضوء الشمس الذي يخرج منه في وقت كسُوفها يكون ¬omF على مثل شكل الثقب، إذا كان السطح الذي يظهر عليه الضوء موازيًا لسطح الثقب. BO‖P

⌐omF وأما ضوء القمر فإنه إذا خرج من ⌐omL ثقب من الثقوب وظهر على سطح موازي للثقب وكان الثقب مستديرًا، فليس يكون الضوء إلا مستديرًا وان كان الثقب ضيقًا ¬omF كان القمر هلالاً أو كان ممتلئًا أو فيما بين ⌐omF ذلك في أول الشهر كان ذلك أو في آخره أو في وقت كسوفه. وكذلك إذا خرج ضوء القمر من ¬omL الثقوب المختلفة ¬omF الأشكال إنما يكون صورها على مثل أشكال الثقوب. BO‖P

---

28 أضواء] اضوا FBPOL || 28 تخرج] تخرج F يخرج OL s.p. BP the verb تخرج follows the maǧāzī fem sing الشمس
|| 29 الشمس] om F || 29 الثقب الواسع] om F || 30 ظهر الضوء الذي يخرج منه] 5 words om F || 31 الثقب
يكون وقت كسُوفها ... مربعًا ظهر] 26 words om F || 31 ظهر الضوء الذي يحرج منه مربعا] marg F || 32 ضوء الشمس
الذي] الضوء L || 35 وأما ضوء القمر ... وان كان الثقب ضيقًا] 27 words om F || 35 إذا خرج ضوء ... ثقب من الثقوب
القمر من] 48 words om L || 38 ذلك في أول الشهر ... من الثقوب المختلفة] 21 words om F || 39 صورها] om L ||
40 post الثقوب] وا كانت ضيقه add L

[B82v] Since things are so, we resolved to investigate the cause that makes this effect appear with the Sun and not with the Moon, and why, in the case of the Sun, it happens through narrow apertures and not through wide apertures. |[O] When we carefully investigated the matter and we deeply examined it, its cause was disclosed to us and its reason appeared to us. We then composed this epistle about it. This is when we began our discourse by saying: |[FOP]||[B]

<2. The Principles of the Demonstration>

<2.1 Rectilinear Propagation of Light; Homothety>

[P23r] From every point of every self-luminous body, light radiates in every straight line that can be extended from that point. We have explained that, with due proof and experimentation, in the first book of our work on *Optics*.[15] Every self-luminous body [O79v] facing a dense body with an aperture throws light from each of its points to that aperture by means of a cone, whose apex[16] is that point <of the self-luminous body>, and whose base is the aperture. This implies that if the light diffused[17] in that cone is prolonged towards [L10r] its base ands hits the plane of the body parallel to the plane of the aperture, <a light> will appear on that plane, whose form is like that of the aperture, because any plane cutting the body of a cone parallel to the base of the cone produces a section of the envelope surface of the cone similar to the

15. Lit.: "in the first section (*maqāla*) of our book of Optics." This passage is a paraphrase of the *Optics* I, 3.19: "From every part of every self-luminous body, light radiates in every straight line extending from that part" (Sabra 1989, I: 20), a matter which is studied in depth in Chapter 3. Ibn al-Haytham performed various experiments to test this property, especially in Prop. 3.3, dealing with the propagation of light in dusty chambers, and proved it to be true for the light of the Sun, the Moon, the fire and even accidental lights. Interestingly, Prop. 3.11 considers the light of the Sun at the time of the eclipse.

16. Lit.: "head" (رأسه).

17. Lit.: "extended."

B82v ولما كان ذلك كذلك رأينا أن نبحث عن العلة التي من أجلها يظهر هذا المعنى في الشمس ولا يظهر في القمر، ويظهر في الشمس من الثقوب الضيقة ولا يظهر من الثقوب الواسعة. O| ولما أنعمنا البحث ⌐omL عن ذلك واستقصينا النظر فيه انكشفت لنا علته وظهرلنا سببه فألفنا فيه هذه المقالة omL⌐. وهذا حين ابتدأنا بالقول في ذلك: B||FOP|

P23r كل جسم مُضيء فإنه يخرج من كان نقطة منه ضوء على كل خط مستقيم يصح أن يمتد من تلك النقطة. وقد بينا ذلك بالبرهان والاعتبار جميعا في المقالة الأولى من كتابنا في المناظر. فكل جسمٍ مضيء O79v يقابل جسمًا كثيفًا فيه ثقب، فإن كل نقطة من ذلك الجسم المضيء يخرج منها ضوء إلى ذلك الثقب على شكل مخروط رأسه تلك النقطة وقاعدته ذلك الثقب. فيلزم من ذلك أن ذلك الضوء الممتد في ذلك المخروط، إذا امتدّ المخروط في جهة L10r قاعدته وانتهى إلى سطح جسم مواز لسطح الثقب، أن يظهر على ذلك السطح، ضوء شكله شكل ذلك الثقب، لأن كل سطح مستو يقطع جسمًا مخروطًا ويكون ذلك السطح

---

41 ذاك] marg B || 41 نبحث] نبحث FB يبحث OL s.p. P || 41 يظهر هذا] صار هذا F || 43 الثقوب] om L || 43 ولما أنعمنا] فلما انعمنا and الحث عن ذلك واستقصينا 4 words om F || 43 عن ذلك واستقصينا ... فيه هذه المقالة] 16 words om L || 45 ابتدأنا] ابتدأنا FL ابتداينا BPO || 46 كل] وقول add F ante || 46 فانه] انما add F post || 46 يخرج من كان نقطة منه] يخرج منه كان نقطة من transp F || 48 كتابنا في] om F || 50 ضوء] om B || 50 على ذلك] كل F || 51 ₃ ذلك] 3 || 53 مواز] موازي BO || 54 شكل مخروط راسه تلك النقطة وقاعدته ذلك الثقب] marg B || 54 مستو] مستوي BO

base of the cone. |[O] This has been shown in the first chapter of the book of *Conics*.[18] Therefore, what appears from the body of the Sun at the time of the eclipse is crescent-shaped. [P23v] If, at the time of the eclipse, the Sun is facing a dense body with a circular aperture, from each point of the self-luminous crescent,[19] which is the <visible> part of the Sun, a light is emitted to the entire [B83r] area[20] of the aperture and any following surface. It is thus clear that the light, coming from every point of the luminous part of the Sun to the aperture opposite the Sun, is in the form of a cone, whose apex is the point of that luminous part <of the Sun> and the base is the plane of the aperture. All light endlessly extends in straight lines, unless it is intercepted by a dense body. On hitting a dense body, it appears on the surface of that body. The light cone, which comes out from a point of the Sun to the aperture [F118r] opposite the Sun, extends in straight lines to the plane of the body opposite the aperture, and light appears on this plane. If this plane is parallel to the plane of the aperture, the shape of the light that appears on the plane [L10v] parallel to the aperture will be in the form of the aperture. |[P]||[BO]

---

18. This is a reference to Apollonius, *Conics*, I,4: “Si l’une ou l’autre de deux surfaces opposées par le sommet est coupée par un certain plan parallèle au cercle sur lequel se déplace la droite qui décrit la surface, le plan intercepté par la surface sera un cercle ayant son centre sur l’axe” (Decorps-Foulquier et al. 2008: 17). This property was generalized to all sections in *Conics* VI, 26 (hyperbolas) and VI, 27 (ellipses). Although only basic knowledge of conics appears here, Ibn al-Haytham gained—conceivably at a later date—extensive knowledge of Apollonius, as evidenced by his *Completion of the Conics*, which aimed at reconstructing the lost Book VIII (Hogendijk 1985; Rashed 2000: 146–271).

19. Arabic has two words for light: ضوء (*ḍaw’*) refers to the light of the light source *per se*, whereas نور (*nūr*) refers to the effect of the light on surrounding objects. This is roughly the difference between the light emitted and the light received. The word *ḍaw’* corresponds to the Greek “φῶς” and to the Latin “lux,” while *nūr* agrees with the Latin “lumen.” In addition, the Arabic term شعاع (*shu‘ā‘*) refers to the lightbeam or radiation of light, which corresponds to the Greek “ἀκτῖνες” and to the Latin “radius.” Surprisingly enough, there is no difference whatsoever in *On the Shape of the Eclipse*, where the term *ḍaw’* is used throughout in all three situations. In order to dispel ambiguity, I have translated throughout the crescent Sun by “self-luminous crescent” and the patch of light by “light crescent.”

20. Lit.: “surface.”

موازيًا لقاعدة ذلك المخروط، فإن الفصل المشترك بين السطح المستوي وبين سطح المخروط يكون شكله شبيها بشكل قاعدة المخروط. [O]| وقد تبيّن هذا المعنى في المقالة الأولى من كتاب المخروطات. والذي يظهر من جرم الشمس في وقت كسُوفها ⌝[omL] يكون على شكل الهلال. [P23v] فإذا قوبلت الشمس في وقت كسُوفها ⌜[omL] بجسم كثيف فيه ثقب مستدير، فإن كل نقطة من الهلال المضيء، الذي هو جزء من الشمس، يخرج منها ضوء إلى جميع [B83r] سَطح الثقب ⌝[omL] وإلى جميع ما يليه من سطوح الأجسام. فيعرض من ذلك أن يكون الضوء، الذي يخرج من كل نقطة من الجزء المُضيء من الشمس إلى الثقب المقابل للشمس، ⌜[omL] على شكل مخروط، رأسه تلك النقطة من الجزء المُضيء وقاعدته سطح الثقب. وكل ضوء فهو يمتدّ أبدا على استقامة مالم يلق جسمًا كثيفًا. فإذا لقي جسمًا كثيفًا ظهرَ على سطح ذلك الجسم. فالمخروط المضيء، الذي يخرج من نقطة من الشمس إلى الثقب [F118r] المقابل للشمس، يمتد على استقامة إلى أن ينتهي إلى سطح الجسم المقابل للثقب ويظهر الضوء على هذا السطح. وإذا كان هذا السطح موازيًا لسطح الثقب، كان شكل هذا الضوء الذي يظهر على السطح [L10v] الموازي للثقب على مثل شكل الثقب. [OB]||[P]|

---

57 والذي يظهر من جرم] فإذا قومت L || 58 يكون على شكل ... في وقت كسُوفها] 10 words om L || 61 وإلى جميع ما ... الثقب المقابل الشمس] 27 words om L || 64 post النقطة] التي add L || 67 يمتد] s.p. BP the verb يمتد follows the masc.sing. المخروط || 68 سطح] سطح BPL لسطح FO || 69 موازيًا لسطح] موازنًا لسطح B

<2.2 The Point-Analysis of the Image>

It is clear from what we have mentioned [P24r] that if, at the time of the eclipse, the Sun is facing a dense body with a circular aperture, its light emerges from every point of the visible crescent of the Sun to the aperture, extends to the plane parallel to the aperture and forms a circular light on this plane. Therefore, on the plane parallel to the aperture, there will be circular lights, compact and overlapping each other, none of which is distinguished from the rest. The totality of these lights is endlessly surrounded by a continuous shadow, which is the shadow of the dense body surrounding the aperture. ‖[FBO]

Consider[21] now the shape of the perimeter of this <solar crescent-shaped> light and its magnitude. |[FO]

We say that, if from every point of every self-luminous body a light is emitted along any straight line properly extended from that point, then the entire body will be luminous, and the light will be emitted from all of it to any opposite point, provided that there is no dense body between that [B83v] point and the self-luminous body that would cut off some of the imaginary lines which lie between that point and the whole self-luminous body.[22] This implies that each point of the plane of the aperture opposite the Sun at [L11r] the time of the eclipse receives the light cone, whose base is the self-luminous crescent, which is the visible part of the Sun, and whose apex is that point of the aperture. If this cone is prolonged [P24v] on the side of its apex, it will hit the plane parallel to the aperture. Thus is formed another cone, whose apex is

---

21. Lit.: "the shape... is investigated" (passive form).

22. This is reminiscent of Ibn al-Haytham, *Optics* I, 3.1: "We find that the light of every self-luminous body radiates on every body opposite to it when there is not between them an opaque or non-transparent body that screens one from the other" (Sabra 1989, I: 13).

فيتبين مما ذكرناه أن P24r الشمس في وقت كسُوفها، إذا قوبلت بجسمٍ كثيفٍ فيه ثقب مستدير، فإن كل نقطة من الهلال الظاهر من الشمس يخرج منها ضوء إلى الثقب ويمتد إلى السطح الموازي للثقب ويحدث في هذا السطح ضوءًا مستديرًا. فيصير في السطح الموازي للثقب أضواء مستديرة متراصة متداخلة بعضها في بعض لا يتميز أحدها عن الباقية. ويكون جملة هذه الأضواء متناهية يحيط بها ظل مُتصّل، هو ظل الجسم الكثيف المحيط بالثقب. ‖FBO

فليبْحَث الآن عن شكل محيط هذا الضوء وعن مقداره. |FO

⌐omB فنقول إنه إذا كان كل نقطة من كل جسم مضيء يخرج منها ضوء على كل خط مستقيم يصح أن يمتد ¬omL من تلك النقطة، فإن كل الجسم مُضيء فإن الضوء يخرج من جميعه إلى كل نقطة تقابله، إذا لم يكن بين تلك B83v النقطة وبين الجسم المضيء جسم كثيف يقطع شيئًا من الخطوط المتخيلة التي بين تلك النقطة وبين جميع الجسم المضيء. فيلزم من ذلك أن كل نقطة من سطح الثقب المقابل للشمس في L11r وقت كسوفها يخرج اليها مخروط مضيء قاعدته الهلال المضيء، الذي هو الجزء الظاهر من

74 ضوءًا مستديرًا] ضوء مستدير F || 75 عن] من P عن FBOL || 78 فليبْحَث] فلنبْحَث L فليبْحَث FBO || 79 فنقول إنه إذا كان ... يصح أن يمتد] 20 words om B || 80 فان كل جسم ...المضيء] marg P || 80 الجسم] جسم FBPOL || 81 فإن الضوء يخرج] فإن يخرج الضوء transp L || 83 المتخيلة] om L

that point of the aperture, and opposite the first cone.[23] This cone is cut by the plane parallel to the aperture. The plane of the aperture is parallel to the plane of the circle tangent to the first cone along the perimeter of the Sun, or nearly. Between them, namely, the plane of the aperture and the plane of the circle bounding the Sun, there is no perceptible <angular> difference;[24] for the plane parallel to the aperture is parallel to the plane of the circle bounding the Sun, whenever the aperture is opposite the Sun. The surface of this circle is cut by the hidden surface of the Sun along a concave line, which is the perimeter of the hidden part of the Sun within the plane of the circle bounding the Sun. Thus the plane parallel to the aperture is parallel to the plane of the crescent-shaped figure which [O80r] appears to be the perimeter of the luminous part of the Sun. I mean by the surface of the crescent: the surface that is bounded by two arcs in the plane of the circle surrounding the Sun, for this is the crescent-shaped figure perceived by the sense[25] as the perimeter of the luminous part of the Sun. Light, which occurs on the plane parallel to the aperture from the cone

---

23. This passage possibly alludes to Pseudo-Aristotle's *Problemata* XV, 11, reasoning by similarity in opposite cones. The Greek and Latin passages read as follows: "δύο γιίνονται κῶνοι, δ' τ' ἀπὸ τοῦ ἡλίου πρὸς τὴν ὀπὴν καὶ δ' ἐντεῦθεν πρὸς τὴν γῆν, καὶ συγκόρυφοι," "Duo fiunt coni eodem apice utentes, qui de sole ad foramen extenditur, et qui inde ad terram."

24. اختلاف محسوس (*ikhtilāf maḥsūs*) or "significant difference," has the same root as the word "sense"—Latin: "differentia sensibilis." See note 25. Wiedemann translates this passage as: "Der Tangentialkreis des Kegels liegt etwas näher bei uns als der Sonnenmittelpunkt" (1914: 156). This is a mistake: Wiedemann points out the *difference of distance* between the circle tangent to the Sun and the center of the Sun, while Ibn al-Haytham had in mind the *angular difference* between the circle tangent to the Sun and the plane of the aperture.

25. الحس (*al-ḥiss*) here refers to eyesight. The Arabic root *ḥss* is long-winded: *aḥassa* (sentire), *al-ḥass*, *ḥassā*, *ḥiss* and *iḥsās* (sensus), *ḥassās* and *maḥsūs* (sensibilis) and *al-ḥass* (sentiens). *On the Shape of the Eclipse* has no fewer than twelve items of "sense" and its derivatives. They could have been rendered by the French "sens, sensible, insensible" but, given the peculiarities of English, I decided on "perception, perceptible, imperceptible, etc." to avoid any confusion. For a study of Ibn al-Haytham's approach to perception, see Sabra (1978: 169–85; 1989, I: 67–8). For a Latin echoing of this theory (Smith 2001, 1: lii–lxxx).

الشمس، ورأسه تلك النقطة من الثقب. وهذا المخروط إذا امتد من P24v جهة رأسه، فهو ينتهي إلى السطح الموازي للثقب. ويحدث منه مخروط رأسه تلك النقطة من الثقب مقابلا للمخروط الأول. وهذا المخروط ينقطع بالسطح الموازي للثقب. وسطح الثقب موازي لسطح الدائرة التي تماس عليها المخروط الأول محيط الشمس، أو ليس بينه، أعني سطح الثقب، وبين سطح الدائرة المحيطة بالشمس اختلاف محسوس؛ فالسطح الموازي للثقب مواز لسطح الدائرة المحيطة بالشمس، إذا كان الثقب مواجهًا للشمس. وسطح هذه الدائرة يقطع السطح المستتر من الشمس، فالخط المقعر الذي هو محيط الجُزء المستتر من الشمس في سطح الدائرة المحيطة بالشمس. فالسطح الموازي للثقب هو مواز لسطح الهلال الذي O80r يظهرَ محيطًا بالجزء المضيء من الشمس. وأعني بسطح الهلال: السطح الذي يحيط به قوسان وهو في السطح الدائرة المحيطة بالشمس، لأن هذا الهلال هو الذي يدركه الحس محيطًا بالجزء المضيء من الشمس. فالضوء، الذي يحدث في السطح الموازي للثقب من المخروط الذي قاعدته الهلال المضيء من الشمس، موازي لشكل الهلال المضيء

---

87 الموازي] om L || 88 مقابلا] om L || 89 التي تماس] om L || 90 الشمس] om L || 92 مواز] موازي BO || 92 الثقب] الشمس scr del الثقب marg F || 95 مواز] موازي BO || 96 محيطًا] om F || 96 وأعني] أعني FPOL B || 96 بسطح] بسطح FBPL سطح O || 97 السطح₁] om B || 97 يحيط به] om F || 97 في] repet O || 98 الذي يدركه للحس محيطا بالجز المضيء من الشمس. فالضوء] marg P

whose base is the self-luminous crescent of the Sun, is parallel and similar[26] to the self-luminous crescent [P25r] as required in opposite cones. This notion has been shown in the first book [B84r] of *Conics*.[27] |[O]

Each point of the plane of the aperture receives from the self-luminous crescent of the Sun a cone, whose base is the self-luminous crescent and whose apex is that point of the aperture. This cone hits the plane parallel to the aperture and, from it <the aperture>, an opposite cone is produced. Light cast by the opposite cone onto the plane parallel to the aperture has a shape similar to the shape of the self-luminous crescent, and it is luminous too. |[O]

Light, which appears on the plane parallel to the aperture at the time of the solar eclipse, is an aggregate of light crescents,[28] contiguous[29] and overlapping each other, none of which is distinguished from the rest. Light, which appears at the time of the solar eclipse on the plane parallel to the aperture, is a compound of those contiguous and overlapping equal circles, and its perimeter consists of parts of the perimeters [L12r] of the contiguous circles. Also, it is composed of the contiguous crescents overlapping one another, and the perimeter therefore consists of the <outermost parts of the> arcs bounding the crescents. |[P]||[OB]

---

26. Lit.: "is parallel to... and is similar to..."

27. This is again a reference to Apollonius, *Conics* I, 4: "Si l'une ou l'autre de deux surfaces opposées par le sommet est coupée par un certain plan parallèle au cercle sur lequel se déplace la droite qui décrit la surface, le plan intercepté par la surface sera un cercle ayant son centre sur l'axe" (Decorps-Foulquier and Federspiel 2008: 17).

28. About the difference between the "self-luminous" and "light" crescents, see note 19.

29. Lit.: "connected" or "contiguous," perhaps under the influence of Persian.

فهو على شكل الهلال P25r المضيء لأن ذلك يلزم في المخروطات المتقابلة. وقد تبين هذا المعنى في المقالة B84r الأولى من المخروطات. |°

فكل نقطة من سطح الثقب يخرج اليها من الهلال المضيء من الشمس مخروط قاعدته الهلال المضيء ورأسه تلك النقطة من الثقب. وينتهي هذا المخروط إلى السطح الموازي للثقب ويحدث منه مخروط مقابل له. ويحدث من المخروط المقابل ضوء في السطح الموازي للثقب يكون شكله على شكل الهلال المضيء وشبيهًا به. |°

فالضوء، الذي يظهر في السطح الموازي للثقب في وقت كسوف الشمس، هو مركب من أهلة مضيئة متصلة متداخلة بعضها في بعض، لا يتميز أحدهما من الباقية. فالضوء، الذي يظهر في وقت كسوف الشمس على السطح الموازي للثقب، مركب من دوائر مُتصّلة متداخلة متساوية، محيطه مركب من أجزاء من L12r محيطات الدوائر المتصلة. وهو مع ذلك مركب من أهلةٍ متصلة متداخلة، فمحيطه مركب من محيطات القسي المحيطة بالأهلة. |P||OB|

101 post المضيء] من الشمس add F || 110 من] من FBPL عن O

[F118v] Each cone coming from the full self-luminous crescent to one point of the aperture is bounded by two <envelope> surfaces of cone, one convex, the other concave. The convex one is tangent to the Sun [P25v] sphere. Its base is an arc of the perimeter of the circle bounding the Sun sphere, its center is under the center of the Sun and nearby the center of the Sun. The concave surface is tangent to the Moon sphere along an arc of the circle that bounds the Moon sphere. Then, this concave surface of the cone extends to the Sun sphere. The Sun sphere is cut off along an arc of a circle equal to the circle, which is the base of the convex surface, for the cone [B84v] bounding the Sun sphere is equal to the cone bounding the Moon sphere, as shown by the mathematicians. The bases of the two <envelope> surfaces of the cone, the convex and the concave, are two arcs of two equal circles. If we imagine the intersection of the convex surface of the cone [L12v] with the concave surface of the cone, this makes a crescent-shaped figure bounded by two arcs of two equal circles. The line joining the apex of the cone to the center of the base of the convex surface, which is the axis[30] of the convex surface, ends at the center of the Sun. Similarly, the line joining the apex of the cone to the center of the base of the concave surface, which is the axis of the concave surface of the cone, passes through the center of the Moon. |[P]||[OB]

30. Lit.: "arrow" (سهم). This term refers to both the axis and the generatrix of the cone. Here it is the axis of the cone.

F118v وكل مخروط يخرج من جميع الهلال المضيء إلى نقطة من الثقب يحيط به سطحان مخروطان أحدهما محدّب والآخر مقعر. فالمحدب مُماسّ لكُرَة P25v الشمس. وقاعدته قوس من محيط دائرة محيطة بكُرة الشمس مركزها تحت مركز الشمس وقريب من مركز الشمس. والسطح المقعر مماس لكُرة القمر على قوس من دائرة محيطة بكُرة القمر. ثم يمتد هذا السطح المخروط المقعر حتّى ينتهي إلى كُرة الشمس. فيقطع كُرة الشمس على قوس من دائرة مساوية للدائرة التي هي قاعدة السطح المحدب، وذلك أن المخروط B84v الذي يحيط بكرة الشمس مساو للمخروط الذي يحيط بكرة القمر، وقد تبين ذلك أصحاب التعاليم. فالسطحان المخروطان المحدب والمقعر قاعدتاهما قوسان من دائرتين متساويتين. فإذا توهمنا سطح [قاعدة] المخروط المحدب يقطع L12v سطح المخروط المقعر، حدث منهما شكل هلالي يحيط به قوسان من دائرتين متساويتين. فيكون الخط الذي يخرج من رأس المخروط إلى مركز قاعدة السطح المحدب، الذي هو سهم السطح المحدب، ينتهي إلى مركز الشمس. ويكون الخط، الذي يخرج من رأس المخروط إلى مركز قاعدة السطح المقعر، الذي هو سهم السطح المخروط المقعر، يمر بمركز القمر. |P||OB

---

115 نقطة من] من om F نقطتين L || 119 لكُرة1] لكرة O || 122 مساو] مساوي OB || 123 تبين] تبين FO بين FBPOL which is a word too many in this context || 125 قاعدة] repet B || 125 متساويتين] ويتين BL s.p.P || 130 السطح2] om FO || 128 المحدب2] المحدث L || 128 سهم] السهم O || 128 المحدب1] المحدث L || 130 القمر] om F

If the convex surface [P26r] is prolonged on the other side of its apex, which is a point of the plane of the aperture, and reaches the plane parallel to the aperture, it will form an arc of a circle on this plane. |[O]

If the generatrix[31] of the surface of the cone is prolonged up to the plane parallel to the aperture, it will describe an arc on this plane. The convexity of this arc will be on the side opposite[32] the convexity of the self-luminous crescent, which is part of the Sun. Similarly, if the concave surface is prolonged on the other side of its apex and reaches the plane parallel to the aperture, it will form an arc of a circle. If the axis of this cone is prolonged up to the plane parallel to the aperture, it will end at the center of the arc that occurs in this plane. The concavity of the arc will be on the side opposite [L13r] the concavity of the self-luminous crescent. The convex face of these two arcs will be on the side of the concave face [B85r] of the self-luminous crescent; their concave face will be on the side of the convex face of the self-luminous crescent. Therefore these two arcs [O80v] form a crescent-shaped figure similar to the self-luminous crescent, though contrary in position.[33] And the two arcs of this crescent are made of two equal circles, because the arcs of the self-luminous crescent <are> made of two equal circles. |[O]||[P]

31. Lit.: "arrow" (سهم). This is now the "generatrix." See note 30.

32. Lit.: "against, versus."

33. See Commentary, pp. 114–6.

ويكون السطح المحدب إذا P26r امتد من جهة رأسه، الذي هو نقطة من سطح الثقب، وانتهى إلى السطح الموازي للثقب، يحدث في هذا السطح قوسًا من دائرة. |O

وإذا امتد سهم هذا السطح المخروط وانتهى إلى السطح الموازي للثقب، انتهى إلى القوس التي يحدث في هذا السطح. ويكون حدبة هذه القوس في ضدّ الجهة التي تلي حدبة الهلال المُضيء الذي هو جزء من الشمس. ويكون السطح المخروط المقعر، إذا امتدّ في جهة رأسه وانتهى إلى السطح الموازي للثقب، يحدث فيه قوسًا من دائرة. وإذا امتدّ سهم هذا المخروط وانتهى إلى السطح الموازي للثقب، انتهى إلى مركز القوس الذي يحدث في هذا السطح. ويكون تقعير هذه القوس في ضد L13r الجهة التي فيها تقعير الهلال المضيء. ويعرض في هاتين القوسين أن يكون حدبتاهما في B85r جهة تقعير الهلال المضيء؛ وتقعيراهما في جهة تحدب الهلال المضيء، ويصير من هاتين القوسين O80v هلال شبيه بالهلال المضيء ويكون وضعه مخالفًا لوضعه. ويكون قوسا هذا الهلال من دائرتين متساويتين، لأن قوس الهلال المضيء <يكون> من دائرتين متساويتين. |O||P

---

132 المحدب] المحدث L || 135 سهم] سطح L || 135 السطح₁] om FOL || 140 هذا] om FOL || 141 الذي] التي FOL || 143 حدبتاهما] حديثاهما B حدبناهما L || 143 وتقعيراهما في جهة تحديب الهلال المضيء] marg B || 143 تحدب] تحديب OL تقديم F

<2.3 Geometric Construction>

If we imagine [P26v] a cone, whose base is the self-luminous crescent that is part of the Sun and the apex is the center of the aperture, and imagine the cone extended to the plane parallel to the aperture, the plane parallel to the aperture will display a crescent similar to the self-luminous crescent. The endpoints of the axes of the two conic surfaces that bound the cone will be the centers of its arcs.[34] We imagine a straight line between the two tips of the crescent. We divide it in two halves, and we produce a perpendicular line through the midpoint. This <line> will go through the centers of the two arcs bounding the crescent. This vertical line and the axes of the two <envelope> surfaces of the cone are in one plane. This plane will cut across the plane of the aperture, forming there a diameter parallel to the vertical line that passes through the centers [L13v] of the two arcs bounding the crescent on the plane parallel to the aperture. This plane will cut the self-luminous crescent, which is part of the Sun, and will pass through the centers of the two arcs that bound it. |[FO]||[BP]

34. This is the cone whose directrix is crescent-shaped.

وإذا توهمنا P26v مخروطًا قاعدته الهلال المضيء الذي هو جزء من الشمس ورأسه مركز الثقب، وتوهمنا المخروط ممتدًا حتّى ينتهي إلى السطح الموازي للثقب، حدث في السطح الموازي للثقب هلال شبيه بالهلال المضيء. ويكون مركزا قوسيهِ طرفي سهمي السطحين المخروطين المحيطين بهذا المخروط. فإذا توهمنا خطًا مُستقيمًا يصل بين طرفي هذا الهلال. وقسمناهُ بنصفين وأخرجنا من منْتصفه عمودًا، فانهُ يمُر بمركزي القوسين المحيطين بالهلال. ويصير هذا العمود وسهما السطحين المخروطين في سطح واحد. وهذا السطح يقطع سطح الثقب، فهو يحدث فيه قُطرًا موازيًا للعمود، الذي يمر بمركزي L13v قوسي الهلال الذي في السطح الموازي للثقب. وهذا السطح يقطع الهلال المضيء الذي هو جزء من الشمس ويمر بمركزي القوسين المحيطين به. |FO||BP|

148 المضيء] om L || 151 قوسيهُ] قوسيه FP قوسية O قوسة B قوسين L || 156 موازيًا] موازنًا B || 156 للعمود] العمودي L || 158 قوسا به يحيط الذي هو ... القوسين بمركزي ويمر] marg P

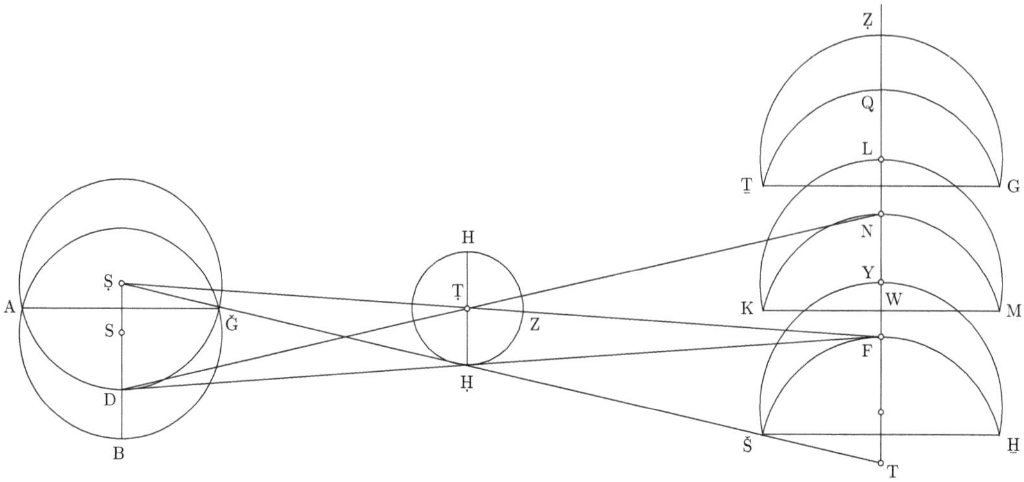

Diagram 1[35]

<Diagram 1> Let the self-luminous crescent, which is part of the Sun, be bounded by the arcs ABǦ and ADǦ, and let the line joining its two tips be AǦ. Let the center of the arc ABǦ be point S, and let the center of the arc ADǦ be point Ṣ. Let the aperture be the circle HḤ of center Ṭ. The crescent, which is received on the plane parallel to the aperture and bounded by the cone whose base is the [P27r] self-luminous crescent and the apex is the center of the aperture, [B85v] is that crescent bounded by the arcs KLM KNM. Let the chord joining its two tips be line KM, its midpoint W, and let the perpendicular drawn through point W down to point T, be line NL. Thereby the centers of the arcs KNM KLM are on the line WT. Let the

35. Diagram 1 depicts the rays of the Sun ABǦ, partially eclipsed by the Moon ADǦ, that enter the aperture HZḤ and produce upside down images beyond: ŠYḪF is obtained through point Ḥ, KLMN through point Ṭ.

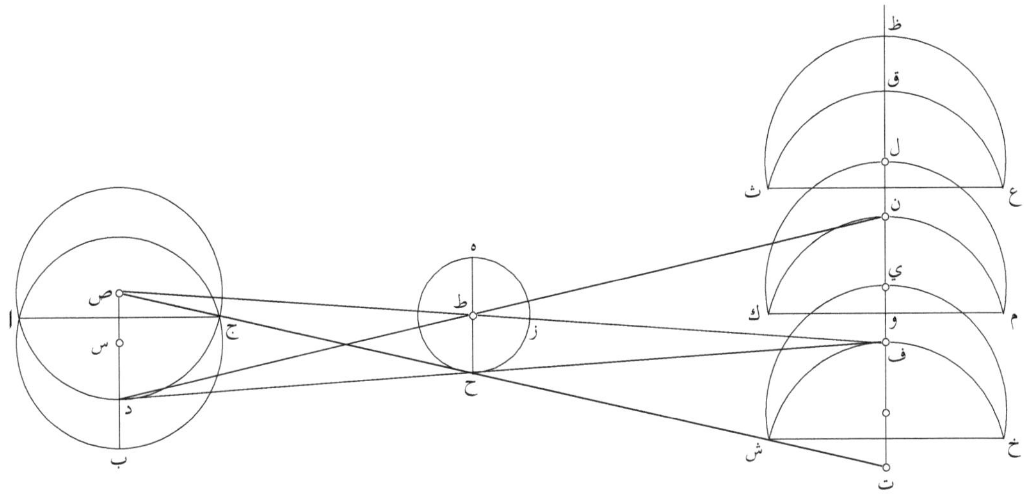

<شكل ١>

<Fig.1> فليكن الهلال المُضيء، الذي هو جزء من الشمس، هو الذي يحيط به قوسا ابج ادج والخط الواصل بين طرفيه اج. وليكن مركز قوس ابج نقطة س ومركز قوس ادج نقطة ص. وليكن الثقب ەح وليكن مستديرًا وليكن مركزهُ ط. وليكن الهلال الذي يحصل في السطح الموازي للثقب الذي تحد به المخروط الذي قاعدته الهلال P27r المضيء ورأسه مركز الثقب، هو B85v الهلال الذي يحيط به قوسا كلم كنم. والخط الواصل بين طرفيه خط كم، ووسطه نقطة و والعمود الذي يخرج من نقطة و عمود ونل ولننفذه في جهة و إلى نقطة ت. فيكون مركزا قوسي كنم كلم على خط وت. وليكن مركز قوس كنم نقطة ف ومركز قوس كلم

161 والخط الواصل بين طرفيه اج] marg P || 161 اج وليكن مركز قوس] marg B || 164 تحد به] يحد ثه L || 167 ولننفذه] ولينفذه B ولننفذه FPOL || 167 نقطة ت] نقطة ن PO || 168 نقطة ي] ومركز قوس كلم marg B

center of arc KNM be point F, and the center of arc KLM, [F119r] be point Y. Then points F Y are the endpoints of the axes of the two surfaces of the cone that emerge from the two arcs ADĞ ABĞ, and whose apex is the point Ṭ, which is the center of the aperture. [L14r] Let the diameter of the aperture, which is made in the plane to which the line WT and the axes of the two surfaces of the cone belong, be line ḤṬH. This plane <TWṬ> cuts across the plane of the arcs ABĞ ADĞ and goes through their centers <S Ṣ>. Let the intersection of this plane <TWṬ> with the plane of the two arcs <ABĞ ADĞ> be line ṢSDB. We draw ṢṬ and extend it straight on. It will end at point F, which is the center of the arc KNM, because line [P27v] ṬF is the axis of the surface of the cone, whose apex is the point Ṭ and the base is the arc KNM. I mean this cone <KNMṬ> is linked to the <opposite> cone whose apex is the point Ṭ and the base is the arc ADĞ. ||[FPO]

<2.4 Relation Between the Distance and Size of the Aperture>

We draw ṢḤ and extend it straight on; it will intercept the line FT, for the lines FT ṬḤ are parallel, and as point Ṣ is in the same plane, line ṢḤ will end at point T. I say first that: if the ratio of (the semi-diameter of the aperture) to (the semi-diameter of the Sun, which is line DṢ)[36] is equal to, or has no perceptible[37] difference with, the ratio [B86r] of (ṬF, which is the distance between the plane parallel to the aperture and the plane of the aperture) to (FṢ, which is the distance between the plane parallel to the aperture and the Sun), then (line FT, [L14v] which is the distance between the center of the arc KNM and the point T) is equal to (the semi-diameter of the circle KNM, which is line FN).[38]

---

36. Because DṢ, the radius of the Moon, equals the radius of the Sun BS.

37. Lit.: "sensible" (محسوس). See note 25.

38. See Commentary, p. 118.

F119r نقطة ي. فيكون نقطتا ف ي هما طرفا سهمي السطحين المخروطين اللذين يخرجان من قوسي ادج ابج، اللذين رأسهما نقطة ط التي هي مركز الثقب. L14r وليكن قطر الثقب الذي يحدثه السطح الذي فيه خط وت، وسهما السطحين المخروطين خط حطه. وهذا السطح يقطع سطح قوسي ابج ادج ويَمُر بمركزيهما. فليكن الفصل المشترك بين هذا السّطح وبين سطح القوسين خط صسدب. ونصل صط وننفذه على استقامة. فهو ينتهي إلى نقطة ف التي هي مركز قوس كنم، لأن خط P27v طف هو سهم السطح المخروط، الذي رأسه نقطة ط وقاعدته قوس كنم. أعني المخروط المتصل بالمخروط الذي رأسه نقطة ط، وقاعدتهُ قوس ادج. ||FPO

ونصل صح وننفذه على استقامة؛ فهو ينتهي إلى خط فت، لأن خطي فت طح متوازيان ونقطة ص في سطحهما فلينته خط صح إلى نقطة ت. فأقول أولا أنه: إذا كانت نسبة ⟨نصف قطر الثقب⟩ إلى ⟨نصف قطر الشمس، الذي خط دص⟩ مساو له أو ليس بينهما اختلاف محسوس كنسبة B86r ⟨طف الذي هو البعد بين السطح الموازي للثقب وبين سطح الثقب⟩ إلى ⟨فص الذي هو البعد بين السطح الموازي للثقب وبين الشمس⟩ فإن ⟨خط فت L14v الذي هو البعد بين مركز قوس كنم وبين نقطة ت⟩ مساو

169 طرفا] طرفي B FPOL طرفا || 178 خط فت] وت FO || 180 أنه] om BL || 180 نسبة] تشَبة B || 181 دص] دم B || 184 فت] تف FO

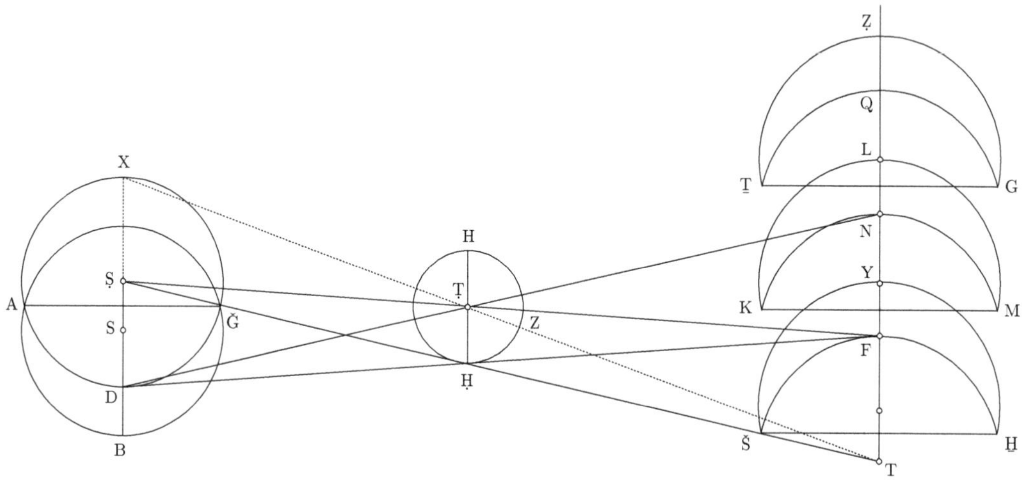

Diagram 1B

<Diagram 1B> If the ratio is such, and line TṬ is drawn and extended straight on to the Sun, it will cut line BṢ beyond point Ṣ <at point X>. [P28r] The ratio of (line ḤT) to (line TṢ) is equal[39] to the ratio (ṬḤ) to (the line <ṢX> that is between point Ṣ and the line going through points T Ṭ, and ends on line BṢ). Thus the line <ṢX> that separates point Ṣ from the line going through points T Ṭ is the semi-diameter of the circle ADĞ. It is equal to line ṢD. The ratio FT to ṢD is composed of the ratio FT to ṬḤ, which is the ratio TṢ to ṢḤ, and of the ratio [O81r] ṬḤ to ṢD, which is the ratio ḤT to TṢ by hypothesis. The ratio ḤT to TṢ composed with ratio TṢ to SḤ is ratio TḤ to ḤṢ. Thus the ratio FT to ṢD is equal to the ratio TḤ to ḤṢ, which is the ratio FṬ to ṬṢ. We join D Ṭ. The line DṬ is on the surface of the cone, whose base is the arc ADĞ and the apex [L15r] is point Ṭ. Line DṬ belongs[40] to the plane of lines ṢD ṢṬ NF, for lines NF [B86v] ṬḤ ṢD are parallel, and line DṬ is in the same plane. If line DṬ is extended straight on, it will end at point N, [P28v] which belongs to

39. I modernize the writing of ratios and proportions by translating مثل and كـ as "equal to."

40. Lit.: "is" (nominal clause).

⟨لنصف قطر دائرة كنم، الذي هو خط فن⟩. وذلك أنه إذا كانت النسبة هذه النسبة ووصلنا خط تط، وأنفذناهُ على استقامة حتّى ينتهي إلى الشمس، فإنه يلقى خط بص خارجًا عن نقطة ص. P28r ويكون نسبة ⟨خط حت⟩ إلى ⟨خط تص⟩ كنسبة ⟨طح⟩ إلى ⟨الخط الذي يفصل بين نقطة ص وبين الخط الذي يمر بنقطتي ت ط وينتهي إلى خط بص⟩. فيكون الخط الذي يفصل بين نقطة ص وبين الخط الذي يمر بنقطتي ت ط، هو نصف قطر دائرة ادج. فهو مساو لخط صد. ونسبة فت إلى صد مؤلفة من نسبة فت إلى طح التي هي نسبة تص إلى صح ومن نسبة طح إلى صد، O81r التي هي نسبة حت إلى تص بالفرض. والنسبة المؤلفة من نسبة حت إلى تص، ومن نسبة تص إلى صح، هي نسبة تح إلى حص. فنسبة فت إلى صد كنسبة تح إلى حص، التي هي نسبة فط إلى طص. ونصل دط. فيكون دط في السطح المخروط الذي قاعدتهُ قوس ادج ورأسه L15r نقطة ط. وخط دط هو في السطح الذي فيه خطوط صد صط نف، لأن خطوط نف B86v طح صد متوازية، وخط دط في سطحها. فإذا خرج خط دط على إستقامة فهو ينتهي إلى نقطة ن، التي هي P28v في السطح المخروط وفي السطح الذي فيه خط نف. فيكون نسبة

---

185 مساو] مساوي BO || 185 قطر] om L || 186 تط] بط B || 187 بص] بدص O || 187 خارجًا] خارجا FL خارجا O || 189 ت ط] طت F || 190 ت ط] طت F || 191 مساو] مساوي BO || 193 صد] صح F || 198 نف طح صد] طن فط صد B || 198 لأن خطوط ... فإذا خرج] marg P || 195 التي] الذي P ||

both the surface of the cone and the plane where line NF is. Thus the ratio NF to ṢD is equal to FṬ to ṬṢ. But the ratio of FT to ṢD is equal to the ratio of FṬ to ṬṢ. Thus line FT is equal to line FN. From this proof, it results that if the ratio of (the semi-diameter of the aperture) to (the semi-diameter of the Sun) is smaller than the ratio of (the distance between the aperture and the plane parallel to the aperture) to (the distance between this plane and the Sun), then line FT, which is the distance between the centers of the two arcs, will be smaller than the semi-diameter of the arc KNM.[41] |[o]

<Conversely> if the ratio of (the semi-diameter of the aperture) to (the semi-diameter of the Sun) is greater than the ratio of (distance) to (distance), then line FT will be greater than the semi-diameter of the circle KNM. It <necessarily> follows that, when the aperture is moved away[42] from the plane that displays the light, or the plane away from it <the aperture>, the line between the two centers[43] will be smaller than the semi-diameter of the arc, because the ratio of distance to distance is then [L15v] greater than the ratio of the semi-diameter of the aperture to the semi-diameter of the Sun. <Conversely,> when the aperture [P29r] is moved closer to the plane <displaying the light>, or the plane closer to it <the aperture>, the line between the two centers will be greater than the semi-diameter of the arc. And if the aperture is moved away from the plane that displays the light, the center of the convex arc will fall inside the concave arc, because the line that bisects[44] the crescent is the same

41. These relationships are further explained in the Commentary, pp. 117 sq.

42. The design of this observation seems to date back to *Problemata Physica* XV, 6: "ἐὰν δὲ πλέον ἀφιστῇ," "si vero amplius removeatur."

43. This is the distance between two crescents $TF' = KS = NF$.

44. Lit.: "is in the middle."

نف إلى صد كنسبة فط إلى طص. وقد كانت نسبة فت إلى صد كنسبة فط إلى طص. فخط فت مثل خط فن. ويلزم من هذا البرهان، أنه إن كانت نسبة ⟨نصف قطر الثقب⟩ إلى ⟨نصف قطر الشمس⟩ أصغر من نسبة ⟨البعد الذي بين الثقب وبين السطح الموازي للثقب⟩ إلى ⟨البعد الذي بين هذا السطح وبين الشمس⟩، فإن خط فت، الذي هو البعد بين مركزي القوسين، يكون أصغر من نصف قطر قوس كنم. I°

فإن كانت نسبة ⟨نصف قطر الثقب⟩ إلى ⟨نصف قطر الشمس⟩ أعظم من نسبة ⟨البعد⟩ إلى ⟨البعد⟩، فإن خط فت يكون أعظم من نصف قطر دائرة كنم. فيلزم من ذلك، أنه متى بَعُدَ الثقب عن السطح الذي يظهر عليه الضوء أو بَعُدَ السطح عنه، أن يصير الخط الذي بين المركزين أصغر من نصف قطر القوس، لأنه يصير نسبة البعد إلى البعد L15v أعظم من نسبة نصف قطر الثقب إلى نصف قطر الشمس. ومتى قرب الثقب P29r من السطح، أو قرب السطح منه، صار الخط الذي بين المركزين أعظم من نصف قطر القوس. فإذا بَعُدَ الثقب من السطح الذي يظهر عليه الضوء، صار مركز القوس المحدبة في داخل القوس المقعرة، لأن الخط الذي في وسط الهلال مساو للخط الذي B87r بين المركزين. وإذا قرب الثقب من

206 كنم] كمم L || 208 قطر] قطره FO || 209 بَعُدَ] بَعُد B بعد FPOL || 210 بَعُدَ] بَعُد B بعد FPOL || 212 نصف] om P || 213 صار] repet del O || 214 بَعُدَ] بَعُد B بعد FPOL || 214 الضوء، صار مركز] marg P || 216 مساو] repet B مساوي O || 216 الثقب من السطح] السطح من الثقب transp F

as the line [B87r] between the two centers. <Conversely,> if the aperture is moved closer to the plane <displaying the light>, the center of the convex arc will fall inside the crescent, on the bisector[45] of the crescent. |[OP]

<2.5 Distinction of the Convex and Concave Faces of the Crescent>

<Diagram 1B> If we imagine a cone whose base is the self-luminous crescent [F119v] bounded by the arcs ABǦ ADǦ, and whose apex is point Ḥ, this cone will be enclosed by the convex surface of the cone whose base is the arc ABǦ, and by the concave surface of the cone whose base is the arc ADǦ. The axis of this concave surface is the line ṢḤ, which ends at point T. If we extend line DḤ, this line will belong to both the concave surface and the plane where are the lines ṢT ṢD FT. If we imagine this cone, whose base is the crescent ABǦD and the apex is point Ḥ, is extended on [L16r] the side of Ḥ, it will hit the plane parallel to the aperture. A crescent [P29v] is bounded by two arcs, one convex, the other concave. The center of the concave arc is point T. The line DḤ ends at a point <F> of the concave arc, which is the intersection between the surface of the cone and the line FT. Let (the line, which is between this point <F> and point T) to (the line DṢ) be equal to the ratio (TḤ) to (ḤṢ). This is the ratio [O81v] FT to DṢ. Line DḤ ends at point F; thus the concave arc passes through point F and the convex arc is <taken> from a circle equal to the circle of

45. Lit.: "the line which is in the middle of the crescent."

السطح، صار مركز القوس المحدبة في داخل الهلال ⌝omB وعلى الخط الذي في وسط الهلال. ⌜omB |OP

وإذا توهمنا مخروطًا قاعدته الهلال المضيء الذي F119v يحيط به قوسا ابج ادج ورأسه نقطة ح، فإن هذا المخروط يحيط به سطح مخروط محدب قاعدته قوس ابج وسطح مخروط مقعر قاعدته قوس ادج. ويكون سهم هذا السطح المقعر هو خط صح الذي ينتهي إلى نقطة ت. وإذا وصلنا خط دح كان هذا الخط في السطح المقعر وفي السطح الذي فيه خطوط صت صد وت. وإذا توهمنا هذا المخروط أعني الذي قاعدته هلال ابجد ورأسه نقطة ح قد امتد في L16r جهة ح، فهو ينتهي إلى السطح الموازي للثقب ويحدث فيه هلالاً P29v يحيط به قوسان محدبة ومقعرة. ويكون مركز القوس المقعرة نقطة ت. وينتهي خط دح إلى النقطة من القوس المقعرة، التي هي الفصل المشترك بين السطح المخروط وبين خط فت. ويكون الخط الذي بين هذه النقطة وبين نقطة ت نسبته إلى خط دص كنسبة تح إلى حص. وهذه النسبة هي نسبة O81v فت إلى دص. فخط دح ينتهي إلى نقطة ف؛ فالقوس المقعرة تمر بنقطة ف، ويكون القوس المحدبة من دائرة مساوية للدائرة التي منها القوس المقعرة

217 وعلى الخط الذي في وسط الهلال] om B || 219 وإذا توهمنا مخروطًا قاعدته الهلال] marg B || 222 إلى] التي F اليه B || 226 ويحدث فيه] وبحد فيه FO || 227 إلى] التي F || 228 الفصل] الفضل L || 229 ت] ث O || 231 ينتهي إلى] marg B || 231 تمر] تمر FB يمر OL the verb follows the masc sing القوس || 232 post ويكون add O مساوي

the concave arc, whose semi-diameter is equal to line FT. Let the concave arc be ŠFḪ. It has been found[46] that line FT equals line FN, and line LY equals line NF, because the crescents KLMN [B87v] ŠYḪF are equal. The two circles that bound them are equal, for they stand at the same distance away from the aperture. The two arcs of each of the two crescents are <made of> two equal circles. |[O]

<3. Analysis of the Image of the Convex Face>

<3.1 Geometric Construction>

<Diagram 1> The convec arc passes through the point Y. Let it be the arc ŠYḪ. If we join points A Ǧ [L16v] to point Ḥ, and extend the two lines [P30r] on the side of Ḥ, they will end at points Š Ḫ. Draw ŠḪ. The ratio ŠḪ to the line AǦ is equal to the ratio of distance to distance, which is the ratio FḤ to ḤD, as well as FṬ to ṬS, and FT to ṢD. Similarly, the ratio KM to AǦ is equal to the ratio of distance to distance, because if we join the two points A Ǧ to point Ṭ, and extend both lines straight on, they will end at points K M. Thus line ŠḪ is equal to KM and parallel to it, because both lines are parallel to line AǦ. ||[BOP]

46. See above, line 202.

فنصف قطرها مساو لخط ف‌ت. فليكن القوس المُقعرة قوس ش‌ف‌خ. وقد تبين أن خط ف‌ت مثل خط ف‌ن، فخط ل‌ي مثل خط ن‌ف، لأن هلالي ك‌ل‌م‌ن B87v ش‌ي‌خ‌ف متساويان. ودوائرهما التّي تحيط بهما متساوية، لأن بُعدهما من الثقب بُعد واحد. و قوسا كل واحدٍ من الهلالين من دائرتين متساويتين. O|

فالقوس المحدبة تمر بنقطة ي. فليكن قوس ش‌ي‌خ. OP| وإذا وصلنا نقطتي ا ج L16v بنقطة ح وأنفذنا P30r الخطين في جهة ح، فإنهما ينتهيان إلى نقطتي ش خ. ونصل ش‌خ. فيكون نسبة ش‌خ إلى خط اج كنسبة البعد إلى البعد، التي هي نسبة ف‌ح إلى ح‌د وكنسبة ف‌ط إلى ط‌ص وكنسبة ف‌ت إلى ص‌د، وكذلك نسبة ك‌م إلى اج هي كنسبة البعد إلى البعد، لأنا إذا وصلنا نقطتي ا ج بنقطة ط وأنفذناهما على استقامةٍ انتهيا إلى نقطتي ك م. فخط ش‌خ مثل خط ك‌م وهو مواز له، لأن كل واحدٍ منهما مواز لخط اج. BOP||

---

233 فليكن] وليكن POL فليكن B || 233 ش‌ف‌خ] ش‌خ‌ف FO || 235 تحيط] تحيط B يحيط FOL the plural دوائرهما requires agreement in fem sing || 238 المحدبة] المحدثه B || 239 ونصل ش خ] ح، فانهما ينتهيان إلى نقطتين ش خ ونصل marg B || 242 إلى البعد] marg F || 244 مواز] موازي BO || 245 مواز] موازي O

Imagine a cone, whose base is the self-luminous crescent and the apex is point H, extended on the side of H. It is a figure <made> upon a part of the crescents of the equal circles of the two convex arcs, that passes through point Y. The arc ŠYḪ is cast on the plane parallel to the aperture and forms a crescent there. This crescent is equal to the first two crescents, because it is as far from the aperture as the first two crescents, or no perceptible[47] difference exists between them. Its distance to the middle crescent is [P30v] equal to the distance from the middle crescent to the last crescent, because the distance from point H to point Ṭ is equal to the distance from point Ḥ to point Ṭ. In this crescent, which is bounded by the arcs ṮẒG ṮQG, the distance from arc ṮẒG to arc KLM is equal to the distance from arc KLM to arc ŠYḪ, and the distance from arc ṮQG to arc KNM is equal to the distance from arc KNM to arc ŠFḪ. Each one of these distances [B88r] is equal to line FT, which is equal to the semi-diameter of any of these arcs, as can be seen and shown in the distance between the first two crescents. |[OP]

Line TL is drawn so that it cuts these two arcs, and it will cut them at points Q Ẓ. We draw line ṮG. It is clear that it is equal to each one of the two lines KM ŠḪ, as it has been shown for lines KM and ŠḪ. If, from any point on arc ADǦ, [F120r] a line is drawn to point Ṭ, it will end at a point on arc KNM. That point on arc KNM is

47. See note 25.

ونتوهم مخروطًا قاعدته الهلال المضيء ورأسه نقطة ه، ونتوهمه ممتدًا في جهة ه. فهو ⌝omFOBL شكلٌ وجزءٌ من الهلالين من دائرتين مساويتين بالقوسين المحدبة يمر بنقطة ي. فليكن قوس شيخ ⌜omFOBL ينتهي إلى السّطح الموازي للثقب ويحدث فيه هلالاً. فيكون هذا الهلال مساويًا للهلال الأول، لأن بُعدهُ من الثقب مثل بُعد الهلالين الأولين، أو ليس بينهما فرق في الحس. ويكون بُعدهُ من الهلال الأوسط P30v مثل بُعد الهلال الأوسط من الهلال الأخير، لأن بُعد نقطة ه من نقطة ط مثل بُعد نقطة ح من نقطة ط. وليكن هذا الهلال، هو الذي يحيط به قوسا ثظع ثقع، فيكون بُعد قوس ثظع من قوس كلم ⌝omL مثل بُعد قوس كلم من قوس شيخ، ويكون بُعد قوس ثقع من قوس كنم ⌜omL مثل بُعد قوس كنم من قوس شفخ. ويكون كل واحدٍ من B88r هذه الأبعاد مثل خط فت الذي هو مساو لنصف قطر كل واحدة من هذه القسي، لأن ذلك يتبَين كما تبين في البعد الذي بين الهلالين الأولين. |OP

فيخرج خط تل حتّى يقطع هاتين القوسين، وليقطعهما على نقطتي ق ظ. ونصل ثع. فيتبين أنه مساو لكل واحد من خطي كم شخ كما تبين

---

247 شكلٌ وجزءٌ من ... شيخ] 14 words om FBOL || 247 وجزءٌ] واخرءًا P || 250 بُعد] بُعدها FO || 250 الأولين] om L || 251 بُعد] بُعدها FO || 252 نقطة₂] om L || 254 مثل بعد قوس كلم] marg P || مثل بُعد قوس كلم ... قوس كنم] 14 words om L || 255 بُعد قوس ثقع ... قوس شفخ] marg P || 255 بُعد₂] om L || 256 واحدٍ] واحدةٍ O || 256 الأبعاد] البعاد scr del الأبعاد marg B || 257 واحدة] واحد FBO || 259 فيخرج] scr del ويخرج marg B || 259 القوسين] repet L || 260 ونصل] ونصل OL ويصل B وتصل F s.p.P || 260 مساو] مساوي BO

the center of the circle, which is the base of the cone whose apex is that point of arc ADĞ and the base is the perimeter of the aperture that gets light from that point of arc ADĞ. [P31r] If a line is drawn from any point on arc ABĞ to point Ṭ, it will end at a point on arc KLM. That point of arc KLM will be the center of the circle, which is the base of the cone, whose apex is that point of arc ABĞ and whose base is the perimeter of the aperture that gets light from that point of [L17v] arc ABĞ. ||[BO]

Consequently, each point of arc KNM is the center of a light circle, whose perimeter falls[48] on the two arcs ŠFḪ ṮQG. Similarly, each point of arc KLM is the center of a light circle, whose perimeter is completed with the two arcs ŠYḪ ṮẒG. ||[O]

Based on what has been established, we say that if light comes from the self-luminous crescent, which is the remaining part of the Sun during the eclipse, [B88v] and faces a dense body with a circular aperture, and the aperture is facing a dense body, light will pass through the circular aperture and will stop at the body opposite the aperture. There it will form a crescent-shaped figure if the plane opposite the aperture is parallel to the aperture and if (the diameter of the aperture) to (the diameter

48. Lit.: "ends up."

في خطي كم شخ. فكل نقطة من قوس ادج F120r إذا خرج منها خط إلى نقطة ط فهو ينتهي إلى نقطة من قوس كنم. ويكون تلك النقطة من قوس كنم مركز الدائرة، التي هي قاعدة المخروط الذي رأسه تلك النقطة من قوس ادج وقاعدته محيط الثقب، التي يخرج إليها الضوء من تلك النقطة من قوس ادج. وكل P31r نقطة من قوس ابج، إذا خرج منها خط إلى نقطة ط، فهو ينتهي إلى نقطة من قوس كلم. ويكون تلك النقطة من قوس كلم مركز الدائرة، التي هي قاعدة المخروط الذي رأسه تلك النقطة من قوس ابج وقاعدته محيط الثقب، التّي يخرج اليها الضوء من تلك النقطة من L17v قوس ابج. ‖BO

فكل نقطة من قوس كنم هي مركز دائرة مضيئة محيطها ينتهي إلى قوسي شفخ ثقع. وكل نقطة من قوس كلم هي مركز دائرة مضيئة محيطها ينتهي إلى قوسي شيخ ثظع. ‖O

وإذ قد تبين ذلك فإنّا نقول أن الضوء الذي يخرج من الهلال المضيء، الذي هو الجُزء من الشمس الباقي بَعدَ المنكسف منها، B88v إذا قوبل بجسم كثيف فيه ثقب مستدير وقوبل الثقب بجسم كثيف، فإن الضوء يخرج من الثقب المستدير وينتهي إلى الجسم المقابل للثقب. ويكون

---

261 خرج] om L ‖ 263 مركز] مركزهُ L ‖ 271 وكل] فكل B ‖ 274 الجُزء] جزء F جز P

[P31v] of the Sun) is equal to the ratio of (the distance that is between the aperture and the plane parallel to the aperture) to (the distance that is between this plane and the Sun). We introduce a lemma for that:

<3.2 A Lemma>

Whatever two equal circles, [O82r] any straight line drawn parallel to the line joining the two centers is equal to the line joining the two centers. |[OP]

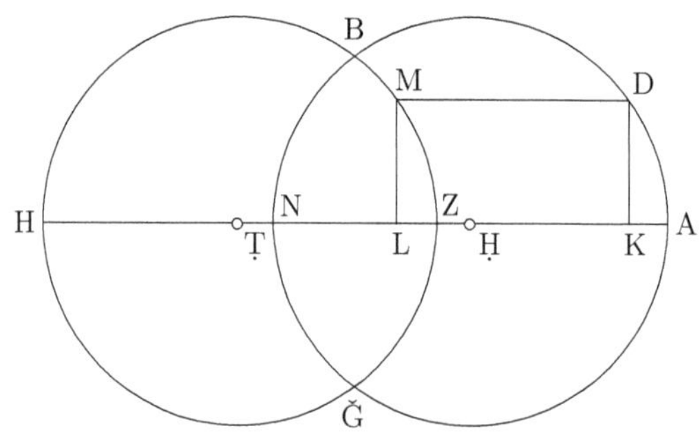

Diagram 2[49]

<Diagram 2> Let the two equal circles be ABĞ BHĞ, with their centers Ḥ Ṭ. We draw ḤṬ and extend it on both sides to A and H. [L18r] It will cut the perimeter of circle ABĞ at points A N and the perimeter of circle BHĞ at points Z H. Take on circle BHĞ any point as agreed, and let it be point M. Line MD is drawn parallel to line AṬ. This line will cut the circle ABĞ at point D. [B89r] From the two points M D two perpendiculars are drawn to line AH; they are DK ML. The two perpendicular

49. The lemma illustrated in Diagram 2 is meant to be applied to the crescent-shaped images ŠYḪF, KLMN and ṮẒGQ in the following Diagram 3.

شكله شكلا هلاليًا، إذا كان السطح المقابل للثقب موازيًا للثقب وكان ⟨قطر الثقب⟩ نسبته إلى ⟨قطر P31v الشمس⟩ كنسبة ⟨البعد الذي بين الثقب وبين السطح الموازي للثقب⟩ إلى ⟨البعد الذي بين هذا السطح وبين الشمس⟩. P32r ولنقدم لذلك مقدمة:

وهي أن كل دائرتين متساويتين O82r يخرج فيما بينهما خط مستقيم مواز للخط الذي يصل بين مركزيهما فإنه مساو للخط الذي يصل بين مركزيهما. |OP

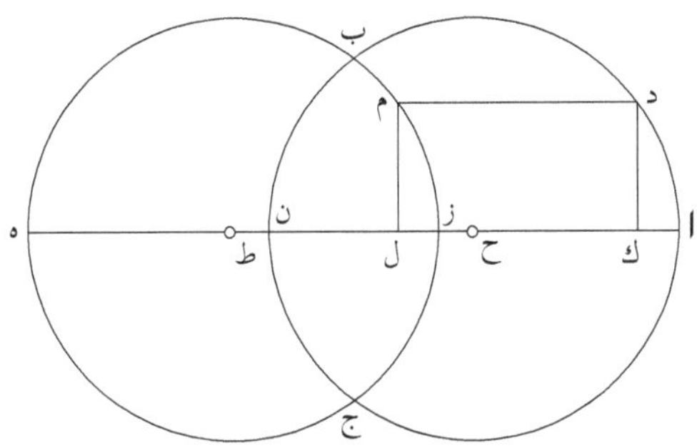

<شكل ٢>

[Fig.2] وليكن دائرتا ابج بهج متساويتين ومركزاهما ح ط. ونصل حط وننفذه في الجهتين إلى ا وإلى ه L18r. وليقطع محيط دائرة ابج على نقطتي ا ن. ويقطع محيط دائرة بهج على نقطتي ز ه. ونفرض على دائرة

278 الذي البعد إلى للثقب الموازي السطح وبين] marg P || 280 لذلك] ذلك B || 281 مواز] موازي BO || 282 مساو] مساوي BO || 285 ونصل ح ط.] om B by homoioteleuton || 286 وننفذه] وينفذه BO وننفذه FL || 288 ما] om BL || 288 ويخرج] وتخرج FL ونخرج O s.p.BP || 288 خط] om P

lines are equal, because the two lines DM AH are parallel, and because the two circles are equal. It follows that arcs AD ZM are equal and lines AK ZL are equal. We thus get KZ, the intersection. KL is equal to AZ, and KL is equal to DM. Then AZ is equal to DM, and AḤ is equal to [P32v] ZṬ, with ḤZ, the intersection. Then AZ is equal to ḤṬ and AZ is equal to DM. Therefore, line DM is equal to line ḤṬ, which is between the two centers. That is what we wanted to demonstrate. |[FPO]

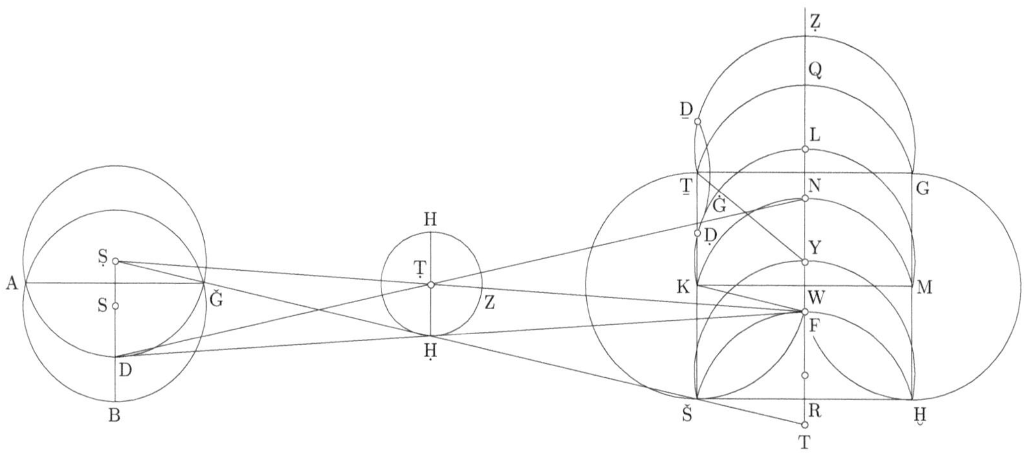

Diagram 3[50]

50. Diagram 3 reproduces Diagram 1, with the addition of arcs ḌḎ, FŠṮ, FḪG. MS B has a marginal addition that aims at criticizing the wrong lines of Diagram 3:

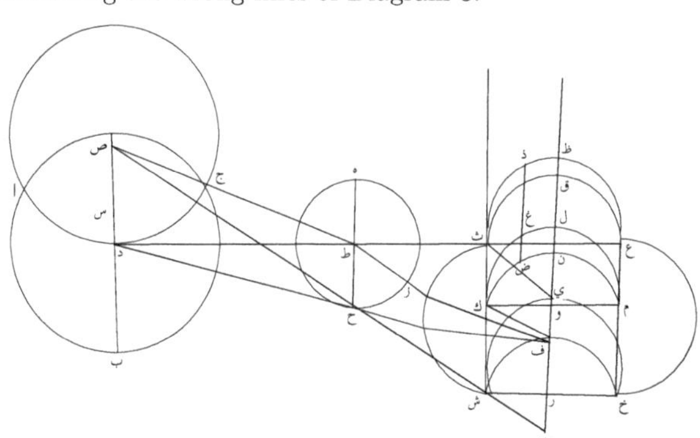

Diagram 3B

"Line ṢĞF is straight as well as DḤF is straight. And we seek to draw a line from A to Ṯ that does not touch DṬ. DṬ could be tangent to the [circular] segment KNM at point N" (lectio incerta).

ب‌ه‌ج نقطة كيف ما اتفق، وليكن نقطة م. ويخرج خط م‌د موازيا لخط اط. فهذا الخط يقطع دائرة اب‌ج فليقطعها على نقطة د. B89r ونخرج من نقطتي م د عمودين على خط اه وليكونا دك م‌ل. فيكون هذان العمودان متساويين، لأن خطين دم اه متوازيان، ولأن الدائرتين متساويتان. يكون قوسا اد زم متساويتين ويكون خط اك زل متساويتين. ونأخذ كز مشتركًا. فيكون كل مثل از و كل مثل دم فـ از مثل دم فـ اح مثل P32v زط وحز مشترك. فـ از مثل حط واز مثل دم فخط دم مثل خط حط الذي بين المركزين. وذلك ما أردنا أن نبين. FPO|

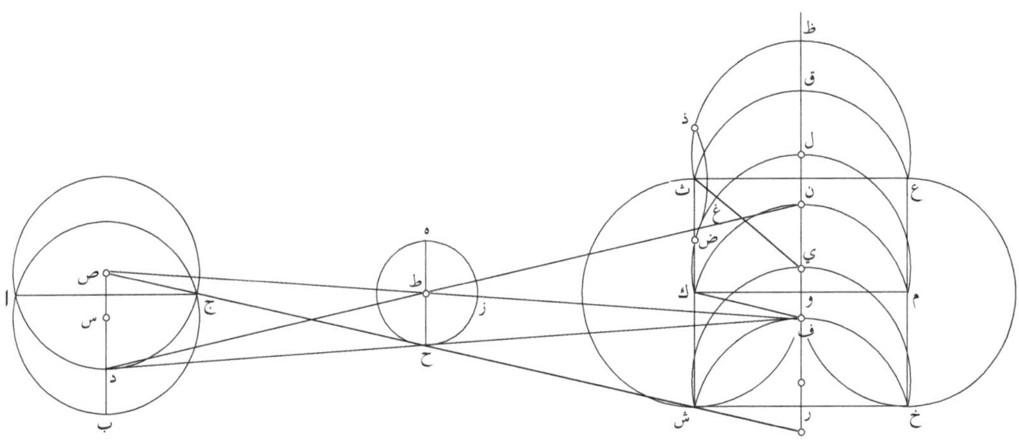

<شكل ٣>

---

289 دائرة] om F || 291 متساويين] متساويتين BL || 291 لأن خطين دم اه متوازيان ولأن الدائرتين متساويتان] marg B || 291 متوازيان] متوازيتين BL || 292 ante قوسا] خط scr del O || 293 لكل 1] لكمل L || 296 B fol.96r add marg 27 words, criticizing the wrong lines of Diagram 3: [ت] خط صجف مستقيم وكذلك دحف مستقيم. ونسعى ان نخرج من ع إلى ث خط ليس على اتصال دط. يكون قد يكون دط مماسا لقطعة كنم على نقطه ن lectio incerta

<3.3 Application of the Lemma>

<Diagram 3> Once this is established, let us go back to the first figure, where it was found that lines ŠḪ KM ṮG are equal and parallel, and line TẒ is perpendicular to line KM, and thus perpendicular to each one of the two lines ŠḪ ṮG. <Line> TẒ passes through the centers of all the arcs, cutting each one of the two lines ŠḪ ṮG. We bisect ŠḪ at point R and get KŠ. It is parallel [L18v] [P33r] to line WR, for KW is equal to ŠR and parallel to it. Then line KŠ is parallel to line NF and equal to it, for KŠ is equal to line FT, which lies between the two centers. And FT is equal to FN. Then line KŠ is equal to line NF, and this holds true of each point of arc KNM. The line drawn up to arc ŠFḪ is parallel to line NF, and equal [F120v] to line NF. The same applies on the two arcs KLM ŠYḪ, on the two arcs ṮQG KNM and on the two arcs ṮẒG KLM. In other words, any line drawn [B89v] between two of these arcs, parallel to the line that bisects the two arcs and passes through their centers, is equal to the line between the midpoints of the two arcs, and is equal to the line between their centers. We draw KṮ. It appears, as we have shown of KŠ, that KṮ is equal to QN and parallel to it. Then line KṮ is equal to line KŠ and aligned[51] with it, for they lie on both sides of one point, which is point K, and are both parallel to one line, which is line TẒ. Thus ŠKṮ form [P33v] a straight line, divided in two halves at point K. Like-

51. Lit.: "connected straight on..."

<Fig.3> وإذ قد تبيّن ذلك فلِنَعُد [إلى] الشكل الأول وقد تبين أن خطوط شخ كم ثع متساوية متوازية، وخط تظ عمود على خط كم فهو عمود على كل واحد من خطي شخ ثع. وتط يمر بمراكز جميع القسي، فهو يقسم كل واحد من خطي شخ ثع بنصفين. ولنقسم شخ على نقطة ر ونصل كش. فيكون موازيًا L18v P33r لخط ور، لأن كو مساو لـ شر ومواز له. فخط كش مواز لخط نف ومساو له، لأن كش مساو لخط فت الذي بين المركزين. وفت مساو لفن. فخط كش مساو لخط نف، وكل نقطة تفرض على قوس كنم. ويخرج منها خط إلى قوس شفخ موازيًا لخط نف، فانه يكون مساويًا F120v لخط نف. وكذلك قوسا كلم شيخ، وكذلك قوسا ثقع كنم وقوسا ثظع كلم. أعني أن كل خط يخرج فيما B89v بين قوسين من هذه القسي، ويكون موازيا للخط الذي يقطع وسطي القوسين ويمر بمركزيهما، فهو مساو للخط الذي فيما بين وسطي القوسين ومساو للخط الذي بين مركزيهما. ونصل كث. فيتبين كما تبين في كش، أن كث مساو لقن ومواز له. فخط كث مساو لخط كش وهو مُتصّل به على استقامة، لأنهما خارجان من نقطة واحِدَة وهي نقطة

297 فلِنَعُد] فلِنَعُد B فلنعد FO || 299 يمر] يمر FL تمر BO the verb يمر follows the masc sing الخط || 300 يقسم] يقسم FBPL ينقسم O the verb يقسم follows the masc sing هو || 300 ولنقسم] ولنقسم BOL ونقسم P وليقسم F || 301 مساو] in the next three lines BO have مساوي and موازي || 302 لأن كش 302 مواز لخط نف ومساو له،] marg B || 302 كش₂] خط add F ante || 304 تفرض] تفرض OL يفرض FB s.p.P the verb تفرض follows the fem sing نقطة || 306 شيخ] شخي FPOL || 308 مساو] مساوي OB || 311 خارجان] خارجا O

wise, if we join MḪ MG by a straight line, and bisect it [L19r] at point M, then take K as a center, and KŠ as a circle opening, the circle will pass through point Ṯ. Likewise, if we take M as a center, and MḪ as a circle opening, the circle will pass through point G. Therefore, circle ŠṮ will cut the circle ṮẒG, for circle ŠṮ is tangent to line ṮG, for KṮ is perpendicular to ṮG, and ṮG cuts the circle ṮẒG. Therefore, circle ŠṮ cuts circle ṮẒG, thus forming an angle at point Ṯ between the two circles. Likewise, circle ḪG cuts circle ṮẒG, thus forming an angle at point G. We draw YṮ. It cuts arc LK because point Y is the center of arc KLM, and cuts it at point Ġ.

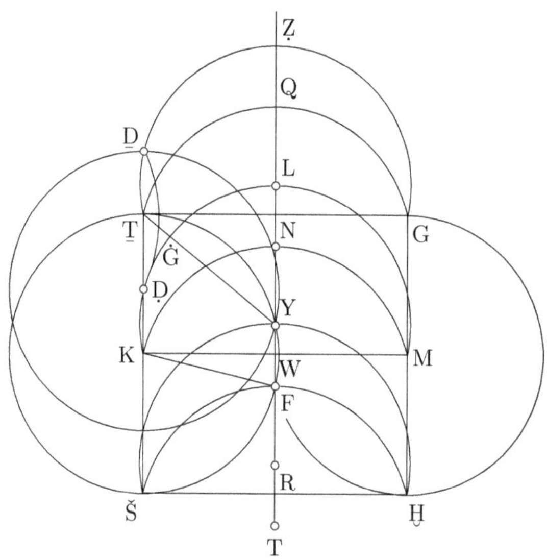

Diagram 3C

<Diagram 3C> We take onto the arc ĠK, a point between the two points Ġ K, close to point K. Let it be point Ḍ. The line joining the two points Ḍ Ṯ is smaller than line KṮ, because it is closer to line ṮĠ. Line ḌḎ is parallel to [P34r] line KṮ. It will be equal to line KṮ, because it is equal to line LẒ. [B90r] Line ḌḎ is greater than the line joining the two points Ḍ Ṯ. If point Ḍ is taken as a center, and a circle is drawn with opening ḌḎ, the perimeter of the circle will pass [L19v] outside [O82v] the

ك، وموازيان لخط واحدٍ وهو خط تظ. فخط شكث متصل P33v على استقامة ومنقسم بنصفين على نقطة ك. وكذلك إذا وصلنا مخ مع اتصلا على استقامة وانقسمَا بنصفين L19r على نقطة م، فنجعل ك مركزًا وندير ببُعد ⌜omB كش دائرة فهي تمر بنقطة ث. وكذلك نجعل م مركزًا وندير ببُعد omB⌝ مخ، دائرة فهي تمر بنقطة ع. فيكون دائرة شث مقاطعة الدائرة ثظع، لأن دائرة شث تماس خط ثع، لأن كث عمود على ثع وثع يقطع دائرة ثظع. ⌜omF فدائرة شث تقطع دائرة ثظع، omF⌝ فيحصل عند نقطة ث زاوية ما بين الدائرتين. وكذلك دائرة خع تقطع دائرة ثظع، ويحصل عند نقطة ع زاوية ما. ونصل يث. فهو يقطع قوس لك لأن نقطة ي مركز قوس ملك، فليقطعه على نقطة غ. ونفرض على قوس غك نقطة فيما بين نقطتي غ ك وقريبة من نقطة ك. وليكن نقطة ض. فيكون الخط الواصل بين نقطتي ض ث أصغر من خط كث، لأنه أقرب إلى خط ثغ. ويخرج ضذ موازيا P34r لخط كث. فيكون مسَاويا لخط كث، لأنه يكون مُساويا لخط لظ. B90r فخط ضذ أعظم من الخط الذي يصل بين نقطتي ض ث. فإذا جُعِل نقطة ض مركزًا وأُدير ببُعد ضذ دائرة، كان محيط الدائرة L19v خارجا O82v عن نقطة ث. فالدائرة التّي مركزها ض ونصف

---

313 مخ مع] صخ مع P || 315-6 كش دائرة فهي ... ببُعد مخ] 12 words om B || 315 ante م] نقطة add L || 316 فيكون] لأن B || 316 مقاطعة الدائرة ثظع لأن دائرة شث] marg B || 318 فدائرة شث تقطع دائرة ثظع] om F || 318 تقطع] تقطع B يقطع FOL s.p.P the verb follows the fem sing دائرة || 319 يقطع FOL s.p.BP the same as above || 325 ضذ] د om P

point Ṯ. The circle, whose center is Ḍ and semi-diameter ḌḎ, surrounds the angle that is at point Ṯ. Light from this circle will cover the angle at point Ṯ, as any other <circle from a> point of arc KLM would do, if it is close to point K. The perimeter of the circle, whose center is that point, passes outside point Ṯ and covers angle Ṯ. The light circles, whose centers belong[52] to the arc KLM, are compact and interlocked[53] with each other. These circles are all equal. The semi-diameter of each one of these circles ends on arc ṮẒG, and full diameter ends on arc ŠYḪ. The perimeter of each one of these circles cuts arc ṮẒG. Some part of their perimeters[54] [P34v] stand outside the arc, following point Ṯ, and some part inside the arc ṮẒG, because every perpendicular drawn from a point of arc ṮẒG to the diameter ṮL is at right angles to the endpoint of the diameter of the circle, whose center is on arc KLM. The endpoint of the diameter is the point where the perpendicular is raised. This circle will be tangent [L20r] to the perpendicular. That part of this circle [B90v] following line ṮL will be inside circle ṮẒG, and that part of this circle following point Ṯ will be partly located outside the circle, because if the above-mentioned perpendicular is extended on the side of point Ṯ, it will pass outside the circle. |°

52. Lit.: "are" (nominal clause).

53. Lit.: "connected with each other."

54. Lit.: "its perimeter."

قطرها ضذ تحيط بالزاوية التي عند نقطة ث. فالضوء الذي يكون في هذه الدائرة يغطي الزاوية التي عند نقطة ث، وكذلك كل نقطة من قوس كلم، إذا كانت قريبة من نقطة ك. فإن الدائرة، التي يكون مركزها تلك النقطة، فإن محيطها يكون خارجًا عن نقطة ث وهي تغطي زاوية ث. والدوائر المضيئة، التي مراكزها على قوس كلم، متراصة متصل بعضها ببعض. وهذه الدوائر كلها متساوية. وكل واحدة من هذِه الدوائر نصف قطرها ينتهي إلى قوس ثظع وتمام قطرها ينتهي إلى قوس شيخ. محيط كل واحدة من هذه الدوائر يقطع قوس ثظع. ويكون بعض محيطها P34v خارجا من القوس مما يلي نقطة ث، وبعضها في داخل قوس ثظع، وذلك أن كل عمود يخرج من نقطة من قوس ثظع على قطر ظل، يكون قائما على طرف قطر الدائرة التي مركزها على قوس كلم. وطرف قطرها النقطة التي خرج منها العمود. فهذِه الدائرة تكون L20r مماسة للعمود. فما كان من هذه الدائرة B90v يلي خط ظل، فهو داخل دائرة ثظع، وما كان من هذه الدائرة يلي نقطة ث فبعضه يقع خارج الدائرة، لأن العمود الذي تقدم ذكره، إذا امتد في جهة نقطة ث حصل خارج الدائرة. I°

---

328 تحيط] يحيط O تحيط FB the verb follows the fem sing الدائرة || 328 post عند] ها add P || 329 يغطي] يغطي F
نغطي PL تغطي B يعطي O the verb follows the masc sing الضوء || 330 نقطة ك] ك om L || 331 تغطي] تغطي FB يعطي
PL تعطي O the verb follows the fem sing هي || 332 متصل] متصل FPOL متراصة scr del متصل corr B || 339 تكون]
تكون F يكون OL om B the verb follows the fem sing الدائرة || 340 ثظع] ظثع transp F || 342 في] من P

Therefore each point of arc ṮẒ is crossed by a light circle,[55] whose center is on arc KL, and some part of which remains outside arc ṮẒ until the center reaches point L, <when> the circle, whose center is point L, will cover the <entire> arc ṮẒG. Similarly, every point of arc ẒG is crossed by a light circle, whose center is on arc LM, and some part of which remains outside arc ẒG. [P35r] What is close to it from the side of point G covers the angle at point G. All these circles are interlocked <with one another>. There is no gap between them, <so that> the interlocking of these circles form a <single> round light from point Š to point Ẓ, and similarly from point G to point Ḫ. ||°

The perimeters of all these circles should end on arc ŠYḪ, because the parallel lines, which are the semi-diameters of these circles, end on arc ŠYḪ such as line KŠ, [L20v] and because the lights are interlocked <with one another> from arc ṮẒG to arc ŠYḪ. Therefore the perimeter of light that appears on the plane parallel to the aperture is convex and circular. ||°

<3.4 The Archimedean Analysis>[56]

[B91r] To demonstrate this notion otherwise, the self-luminous crescent, which is the remaining part of the Sun, <sends>[57] to every point of the perimeter of the aperture a light in the form of a cone, whose base is the light crescent and the apex is that point of the perimeter of the aperture. Each one of these cones extends from that

55. Lit.: "a light circle passes through each point of arc ṮẒ..."

56. See Commentary, pp. 90–1 and 132–4.

57. Reordering the sentence: "that self-luminous crescent... comes out of it... a light..."

فكل نقطة من قوس ثظ تمر بها دائرة مضيئة مركزها على قوس كل، وبعضها خارج عن قوس ثظ إلى أن ينتهي المركز إلى نقطة ل، فيكون الدائرة التي مركزها نقطة ل منطبقة على قوس ثظع. وكذلك كل نقطة من قوس ظع تمر بها دائرة مضيئة مركزها على قوس لم. فيكون بعضها خارجًا عن قوس P35r ظع. ويكون ما يقرب منها من نقطة ع تغطي الزاوية التي عند نقطة ع. وهذه الدوائر متصلة. وليس فيما بينها فرج، ويعرض من اتصال هذه الدوائر أن يكون الضوء مستديرًا من نقطة ش إلى نقطة ظ، وكذلك إلى نقطة ع وإلى نقطة خ. °||

ويكون جميع هذه الدوائر تنتهي محيطاتها إلى قوس شيخ، لأن الخطوط المتوازية التي هي أنصاف أقطار هذه الدوائر تنتهي إلى قوس شيخ كمثل خط كش. L20v فيتصل الأضواء من قوس ثظغ إلى قوس شيخ. فمحيط الضوء الذي يظهر على السطح الموازي للثقب، يكون محدبه مُستديرًا. °||

B91r ولنبين هذا المعنى على وجه آخر، وهو أن الهلال المضيء، الذي هو الجزء الباقي من الشمس، يخرج منه إلى كل نقطة من محيط الثقب، ضوء

---

344 المركز] المراكز BL || 346 تمر بها] تمر بها marg B || 346 تمر] تمر FB يمر OL s.p.P the verb follows the fem sing نقطة || 347 تغطي] تغطي BPL يعطي FO the verb follows the fem sing منها || 351 تنتهي] تنتهي B ينتهي FOL the plural المحيطات requires agreement in fem sing || 352 تنتهي] ينتهي FOL s.p.PB the same as above || 353 فيتصل] فيتصل OL فتتصل B نتصل F s.p.P || 357 إلى كل] إلى FO من كل P

point [F121r] and ends at the plane parallel to the aperture, thus creating a cone [P35v] opposite the first cone, with its base on the plane parallel to the aperture, and its shape similar to the light crescent-shaped figure. |[O]

We have explained this lightening earlier in this discourse. The cones, which extend from the self-luminous crescent to the perimeter of the aperture, are interlocked and overlap one another. The light, which is on the perimeter of the convex <part of the> crescent appearing on the plane parallel to the aperture, is that light coming from arc ABǦ. |[O]

This <result> has been shown too. The convex surface of each cone <that> comes from the self-luminous crescent to a point of the perimeter of the aperture, is the surface whose base is arc ABǦ, [L21r] and the center of this arc is point S. The line that comes out from point S to the point of the perimeter of the aperture, which is the apex of the cone, is the axis of the cone. If the axis is extended to the plane parallel to the aperture, its endpoint will reach the center of the arc bounding [B91v] the crescent, which is in the plane parallel to the aperture. |[P]||[O]

على شكل مخروط قاعدته الهلال المضيء، ورأسه تلك النقطة من محيط الثقب. وكل واحد من هذِه المخروطات يمتد من تلك النقطة F121r وينتهي إلى السّطح الموازي للثقب، ويحدث منه مخروط P35v مقابل المخروط الأول قاعدته في السّطح الموازي للثقب، ويكون شكله شبيها بشكل الهلال المضيء. |°

وقد بينا هذا المضيء فيما تقدم من هذا القول. والمخروطات، التي تخرج من الهلال المضيء إلى محيط الثقب، متصلة متداخلة بعضها في بعض. ويكون الضوء، الذي في محيط محدب الهلال الذي يحدث في السطح الموازي للثقب، هو الضوء الذي يخرج من قوس اب ج. |°

وقد تبين ذلك أيضا. والسطح الحدب من كل مخروط <الذي> يخرج من الهلال المضيء إلى نقطة من محيط الثقب، هو السطح الذي قاعدته قوس اب ج، L21r ومركز هذه القوس هو نقطة س. فالخط الذي يخرج من نقطة س إلى النقطة من محيط الثقب، التي هي رأس المخروط، هو سهم هذا المخروط. وإذا امتد هذا السهم حتّى ينتهي إلى السطح الموازي للثقب، فإن طرفه ينتهي إلى مركز القوس المحيطة B91v بالهلال الذي في السطح الموازي للثقب. °||P|

359 post الثقب] الهلال scr del B || 360 مخروط] مخروطا O || 361 المخروط] om L || 362 بشكل] om B || 364 تخرج] تخرج FB يخرج OL the plural subject والمخروطات requires agreement in fem sing || 364 الثقب] om L || 365 يحدث] om P || 367 أيضا] om P || 368 يخرج] يخرج O تخرج FL s.p.BP the verb follows the masc sing مخروط || 368 من الهلال] om L || 369 قاعدته قوس] قاعدته om L قوس om F

This being the case, the interlocked cones, whose apices are on the perimeter of the aperture, amount to a single cone whose base is the self-luminous crescent [P36r] and its apex can move around the perimeter of the aperture until it returns to its <original> position. If the apex of the cone [O83r] moves[58] along the perimeter of the aperture, its <entire> axis will move along the perimeter of the aperture all together. Thus the endpoint of the axis, which is in the plane parallel to the aperture, will describe a perimeter of a circle parallel to the perimeter of the aperture, provided that the aperture is circular. The full cone of light, which is behind[59] the aperture, moves in <one> movement with the axis. The light crescent, which is on the plane parallel to the aperture, moves along the circle on which the endpoint of the axis moves. The distance of the endpoint of the axis to the convex <part of> the light crescent does not change, because the crescents that appear on the plane parallel to the aperture are equal, [L21v] the distances of these crescents to the perimeter of the aperture are equal, and the distances from the perimeter of the aperture to the Sun are equal. Thus the convex <part of> the crescent, which is on the plane parallel to the aperture, draws on that plane a circle parallel to the circle along which the endpoint of the axis is moving. This circle is the perimeter of the light that appears on the plane parallel to the aperture, that is, the convex perimeter, not the concave, except that the circle is not drawn by only one of the lines in the surface of the light cone. When the apex [P36v] of the light cone is moved along the perimeter of the aperture, every line within the surface of the light cone moves along the perimeter of the aperture all together, and every line of the convex surface of the light cone ends at [B92r] the apex of the cone. Then <on the opposite side> it ends on the convex line of

58. The mechanical reasoning could be strengthened by saying: "If the apex of the cone is driven along the perimeter of the aperture..." see Commentary, pp. 90–1.

59. Lit.: "below."

وإذا كان ذلك كذلك، فالمخروطات المتصلة التي رؤوسها على محيط الثقب هي بمنزلة مخروط واحد قاعدته الهلال المضيء، P36r ورأسه قد تحرك على محيط الثقب حتّى عاد إلى موضعه. وإذا تحرك رأس المخروط على محيط الثقب، فسهمه يتحرك على محيط الثقب. O83r فطرف هذا السهم، الذي هو في السطح الموازي للثقب، يتحرك على محيط دائرة موازية لمحيط الثقب، إذا كان الثقب مستديرًا. وجميع المخروط المضيء، الذي يحت الثقب، يتحرك مع حركة السهم. فالهلال المضيء، الذي في السطح الموازي للثقب، يتحرك حول الدائرة التي يتحرك عليها طرف السهم. وبُعد طرف السهم من محدب الهلال المضيء لا يتغير، لأن الأهلة التي تحدث في السطح الموازي للثقب متساوية، L21v لأن أبعاد هذه الأهلة من محيط الثقب متساوية، وأبعاد محيط الثقب من الشمس متساوية. فمحدب الهلال، الذي في السطح الموازي للثقب، يرسم في هذا السطح دائرة موازية للدائرة التي تتحرك عليها طرف السهم. وهذه الدائرة هي محيط الضوء الذي يظهر في السطح الموازي للثقب، أعني المحيط المحدب دون المقعر، إلا أن هذه الدائرة ليس يرسمها خط واحد من الخطوط التي في سطح المخروط المضيء. وذلك أنه إذا تحرك P36v رأس

---

374 ذلك] om B || 374 رؤوسها] روسها FPL || 376 على محيط الثقب ... حتّى عاد إلى موضعه] marg B || 380 الذي$_1$] om O || 380 يحت] يحت FO تحت B يحب L the verb follows the masc sing المخروط || 382 من] عن L || 382 المضيء] om L || 383 تحدث] تحدث F يحدث OL s.p.BP the plural subject الأهلة requires agreement in fem sing || 384 متساوية] تساوية L || 388 الدائرة] الدوائر F || 389 وذلك أنه إذا تحرك رأس المخروط المضيء] marg B

the crescent on the plane parallel to the aperture; each line moves along the perimeter of the aperture, and the endpoint moves along the perimeter of a circle on the plane parallel to the aperture.

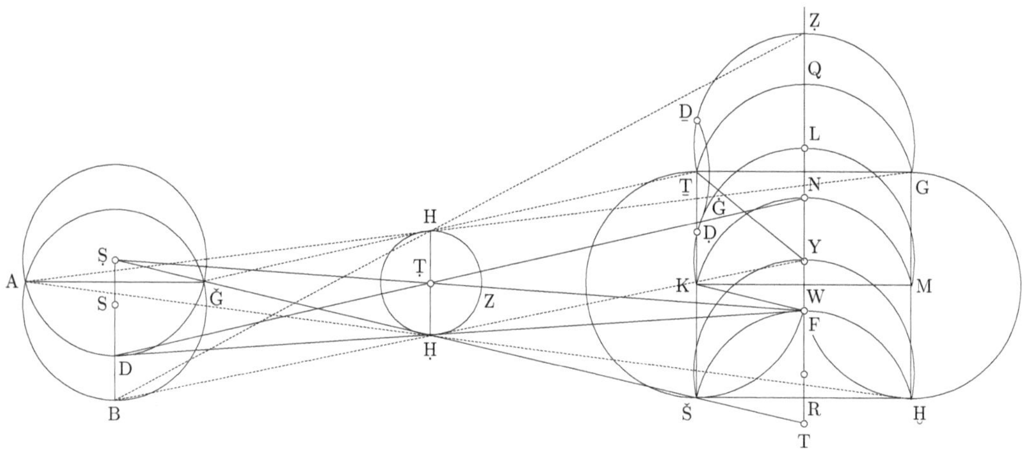

Diagram 3D[60]

<Diagram 3D> The line that comes from point Ğ to [L22r] point Ḥ, ends at point Š. If the apex of the light cone then moves from point Ğ to point B,[61] the endpoint of the line, which is point Š, will move along the perimeter of the circle whose center is point K, because the line that comes from point Ğ to point Ṭ ends at point K, and because this line is the axis of the cone, whose apex is point Ğ and whose base is the perimeter of the aperture. For that reason,[62] the line that comes from point Ğ to point Ḥ ends at point Š. If the apex of the cone moves along the perimeter of the aperture, the endpoint of this line moves along the perimeter of circle ŠT all together, [P37r] until the apex of the cone reaches point H, so that the endpoint of this line

60. In Diagram 3B, Ibn al-Haytham considers the projection of a straight line when its endpoint travels around the entire perimeter of the Sun, from Ğ to B and from B to A, hence the addition of the new dotted lines.

61. Lit.: "... point Ğ, going to point B."

62. ف introduces the conclusion, where the author repeats the statement of line 442.

المخروط المضيء على محيط الثقب، تحرك كل خط في سطح المخروط المضيء على محيط الثقب، وكل خط في السطح المحدب من المخروط المضيء ينتهي إلى B92r رأس المخروط. ثم ينتهي إلى الخط المحدب الذي في الهلال، الذي في السطح الموازي للثقب، وكل خط يتحرك على محيط الثقب، فإن طرفه يتحرك على محيط دائرة في السطح الموازي للثقب. فالخط الذي يخرج من نقطة ج إلى L22r نقطة ح، هو ينتهي إلى نقطة ش. ثم إذا تحرك رأس المخروط المضيء من نقطة ج ذاهِبًا في جهة ب، فإن طرف الخط، الذي على نقطة ش، يتحرك على محيط الدائرة التي مركزها نقطة ك، لأن الخط الذي يخرج من نقطة ج إلى نقطة ط ينتهي إلى نقطة ك، لأن هذا الخط هو سهم المخروط، الذي رأسه نقطة ج وقاعدته محيط الثقب. فالخط الذي يخرج من نقطة ج إلى نقطة ح ينتهي إلى نقطة ش. وإذا تحرك رأس المخروط على محيط الثقب، تحرك طرف هذا الخط على محيط دائرة شت إلى P37r أن ينتهي رأس المخروط إلى نقطة ه. فيكون طرف هذا الخط قد قطع قوس شث وانتهى إلى نقطة ث. ويكون الخط الذي يخرج من نقطة ا إلى نقطة ح ينتهي إلى نقطة خ، ويتحرك حول محيط الثقب بحركة رأس المخروط. فيتحرك طرفه على

393 الهلال] الهلالات corr B scr del || 394 فإن طرفه يتحرك على محيط دائرة] marg B || 395 ج إلى نقطة] om P by homoioteleuton || 398 لأن] ante ذاهِبًا في جهة ب ... مركزها نقطة ك repet del B || 404 نقطة₃] om L || 405 بحركة] بحركة PO يحركة BL s.p.F

can describe the arc ŠṮ and stop at point Ṯ. The line that comes from point A to point Ḥ ends at point Ḫ, and moves along the perimeter of the aperture, driven by[63] the apex of the cone. Its endpoint moves along the perimeter of circle ḪG, though taking the direction of Š. If the apex [B92v] of the cone ends at point H, [L22v] the endpoint of this line will end at point G, so that the crescent ṮẒGQ is completed. Each point of the arc [F121v] ABǦ comes from its line to the apex of the cone and ends at some point of arc ṮẒG. Then, if the apex of the cone moves[64] from point H to point Z, the endpoint of the line coming from point Ǧ to the apex of the cone and hitting point Ṯ, describes the full circle ŠṮ. The fullness of this circle occurs within the light that lies between arc ṮẒG and arc ŠYḪ. And so moves the endpoint of the line that comes from point A to the apex of the cone, that is, the line reaching point G of the arc GḪ, which [P37v] completes[65] the full circle. |P||O

If the apex of the cone moves from point Ḥ to point H, as well as the line coming from point Ǧ to point Š on the perimeter of circle ŠṮ, every line coming from the convex surface, whose base is arc ABǦ and the apex is point Ḥ, moves along the perimeter of the aperture all together. Each one of these lines describes the perimeter of some circle in the plane parallel to the aperture, and [L23r] its center is a point of arc KLM. The endpoint of the line that comes from point Ǧ to the apex of the cone and

63. Lit.: "according to the movement of..."

64. Or, equivalently, "if the apex of the cone is driven from..." see note 58.

65. Lit.: "is the full circle" (nominal clause).

محيط دائرة خع، إلا أنه يكون آخذاً نحو جهة ش. فإذا انتهى رأس B92v المخروط إلى نقطة ه، L22v يكون طرف هذا الخط قد انتهى إلى نقطة ع، ويكون هلال ثظعق قد تم. ويكون كلّ نقطة من قوس F121v ابج يخرج منها خط إلى رأس المخروط وينتهي إلى نقطة من قوس ثظع. ثم إذا تحرك رأس المخروط من نقطة ه نحو نقطة ز، تحرك طرف الخط الذي خرج من نقطة ج إلى رأس المخروط الذي انتهى إلى نقطة ث، على تمام دائرة شث. وتمام هذه الدائرة يكون في داخل الضوء فيما بين قوس ثظع وقوس شيخ. ويتحرك طرف الخط الذي يخرج من نقطة ا إلى رأس المخروط، أعني الخط الذي انتهى إلى نقطة ع على قوس عخ، الذي هو P37v تمام الدائرة. °‖P‖

وإذا تحرك رأس المخروط من نقطة ح إلى نقطة ه، وتحرك الخط الذي يخرج من نقطة ج وينتهي إلى نقطة ش على محيط دائرة شث، تحرك كل خط يخرج في السطح المحدب، الذي قاعدته قوس ابج ورأسه نقطة ح، على محيط الثقب. وكل واحد من هذه الخطوط يتحرك على محيط دائرة في السطح الموازي للثقب، يكون L23r مركزها نقطة من قوس كلم. والخط الذي يخرج من نقطة قوس من ⌐omF نقطة ج إلى رأس المخروط، ينتهي

407 ه] يكون طرف هذا الخط قد انتهى إلى نقطة marg B || 414 الذي] يخرج post scr del انتهى إلى marg P || 416 إلى نقطة ج] نقطة ه وتحرك الخط الذي يخرج من نقطة ج marg B || 416 ح] ه F || 421 نقطة ج إلى رأس ... على دائرة قريبة] 17 words om F

reaches the arc, hits a point near point Š, and moves along a circle near circle ŠṮ, with its center near point K on arc KL. Thus if the endpoint of the line moving on the circle near circle ŠṮ [O83v] to [B93r] a point of arc ṮẒ arrives close to point Ṯ, the arc ṮẒ will be described. |[F]||[BPO]

The same holds for each line within the convex surface, whose base is the arc ABǦ: its endpoint moves on the perimeter of a circle, reaches the arc ṮẒG and describes this arc. ||[O]

[P38r] It has been demonstrated that if the apex of the cone, whose base is the light crescent and <whose> apex is a point of the perimeter of the aperture, moves along the perimeter of the aperture, the convex <part> of the crescent that occurs on the plane parallel to the aperture completes the perimeter of a true circle on this plane through a curve parallel to the circle that is applied to the apex of the cone. The appearance of this circle, namely, that drawn by the convex <part of> the crescent, [L23v] does not <result> solely from the movement of a single line, but from the movements of many shifting lines. Whenever one of them leaves the perimeter of the circle, it is followed by another line that takes its place so as to complement the circle. |[BO]

طرفهُ إلى نقطة قريبة من نقطة ش، ويتحرك على دائرة قريبة ⌜omF⌝ من دائرة شت، يكون مركزها نقطة قريبة من نقطة ك على قوس كل. فإذا صار طرف الخط الذي يتحرك على الدائرة القريبة من دائرة شث O83v إلى B93r نقطة من قوس ثظ قريبة من نقطة ث، قطع قوس ثظ. BPO‖F|

وكذلك جميع الخطوط التي في السطح المحدب الذي قاعدته قوس اب ج: يتحرك طرفهُ على محيط دائرة، وينتهي إلى قوس ثظ ع، ويقطع هذه القوس. O‖

فقد P38r تبين مما ذكرناه، أن المخروط الذي قاعدته الهلال المضيء ورأسه نقطة من محيط الثقب، إذا تحرك رأسه على محيط الثقب، فإن الهلال الذي يحدث في السطح الموازي للثقب، يتمم بحدبه محيط دائرة صحيحة الاستدارة موازية الدائرة، الذي يحدثها سهم المخروط في هذا السطح. ويكون حدوث هذِه الدائرة أعني التي رسمتها حدبة الهلال، L23v لا من حركة خط واحد، بل من حركات خطوط كثيرة متبدلة. كلما خرج واحد منها عن محيط هذه الدائرة صار مكانه خط آخر يليه يتمم الدائرة. BO|

---

423 على قوس كل] three words om B || 425 قطع] ويقطع FBPOL a shared error that must be corrected to complete the sentence || 429 ذكرناه] ذكربا L || 430 محيط ₁] om L || 433 أعني] om F || 435 يليه] يليه B ثليه L lectio incerta

According to our previous two demonstrations, we may conclude that the perimeter of the light, which appears at the time of the solar eclipse on the plane parallel to the aperture, is a proper circular line, <even if> the curve does not achieve a full revolution. And this is what we wanted to show. |°||[FBP]

<4. Analysis of the Image of the Concave Face>

We have now to study <analytically> the image of the concavity that appears in this light. ||[BO]

<Diagram 3D> We say that it has been shown that line ŠK is equal to line FT, which is between the two centers, and equal [P38v] to the line FN, which is between the two arcs. We draw FK. [B93v] FK is equal to FN, because F is the center of arc KNM. Thus line KŠ is equal to line KF, and circle ŠT passes through point F, goes back to point Š, and a part of it falls within arc ŠFḪ. Similarly, every line drawn parallel to line NF from a point of arc KNM, ends on arc ŠFN and is equal to line NF. Every line that comes from a point of arc KNM to point F is equal to line NF. Thus it is equal to the line [L24r] parallel to it, because point F is the center of the arc <KNM>. Therefore the perimeter of each circle, whose center is a point of arc KNM and semi-

فقد تبين مما بيناه بالوجهين جميعا أن محيط الضوء الذي يظهر في وقت كسوف الشمس على السطح الموازي للثقب، هو خط مستدير صحيح الاستدارة ليس بتام الإحاطة. وذلك ما أردنا أن نبيّن. ||FBP ||O

⌐omL وقد بقى أن نبين omL¬ صورة التقعير الذي يظهر في هذا الضوء. ||BO

فنقول أنه قد تبين أن خط شك مساو لخط فت الذي بين المركزين، ومساو P38v لخط فن الذي هو بين القوسين. ونصل فك. B93v فيكون فك متساويا لفن، لأن ف مركز قوس كنم. فخط كش مساو لخط كف، فدائرة شت تمر بنقطة ف وتعود إلى نقطة ش، ويصير جزء منها في داخل قوس شفخ. وكذلك كل خط يخرج من نقطة من قوس كنم. ويكون موازيا لخط نف ينتهي إلى قوس شفن، ويكون مساويا لخط نف. وكل خط يخرج من نقطة من قوس كنم إلى نقطة ف، يكون مساويا لخط ⌐omF نف. فيكون مساويا للخط omF¬ L24r الموازي له، لأن نقطة ف مركز القوس. فكل دائرة مركزها نقطة على قوس كنم ونصف قطرها مساو لخط نف، فإن محيطها يمر بنقطة ف ويقطع قوس شفخ،

436 يظهر] om F || 439 وقد بقى أن نبين] om L || 439 هذا] om F || 440 أن] om L || 440 مساو] in the next three lines, BO have مساوي || 441 هو] om FBOL || 441 ونصل] ونصل FOL ويضل B s.p.P || 441 فيكون فك] marg B || 443 شت] شث BPL || 443 تمر] تمر FO يمر BL the verb follows the fem sing دائرة || 443 وتعود] وتعود FL ويعود BO the verb follows the fem sing دائرة || 447 نف. فيكون مساويا للخط] om F لخط نف. فيكون مساويا repet L || 448 كنم] فنم F || 449 يمر] يمر FBL تمر O the verb follows the masc sing محيط

diameter is equal to line NF, goes through point F and cuts the arc ŠFḪ, and a part of it falls within the concavity of arc ŠFḪ, because every line that comes from point F to the endpoint of the line parallel to line NF, will fall within arc ŠFḪ, thus forming an acute angle with it, except the circle of center N, which is tangent to arc [P39r] ŠFḪ. And all remaining circles, whose centers are on the arc KNM, cut across arc ŠFḪ at two point: one is point F, and the other is the endpoint of the line parallel to line NF. These circles are interlocked, compact and overlap one another from point [B94r] Š to point Ḫ. That of center K cuts arc ŠFḪ on the two points F Š. That of center M cuts arc ŠFḪ on the two points F Ḫ. The remaining circles cut this arc on compact subsequent points, and <cut> the part [L24v] of the circle of center K that falls inside arc ŠF, and most parts that are within this arc. The same applies to the part of the circle of center M, which falls [F122r] within arc FḪ, and most parts that are within arc FḪ. That parts of the remaining circles which are closer to the two outmermost circles, will be greater than the following <ones>. |[BO]

ويحصل جزء منها في داخل تقعير قوس شفخ، لأن كل خط يخرج من نقطة ف إلى طرف الخط الموازي لخط نف يقع في داخل قوس شفخ، ويحيط معه بزاوية حادة، الّا الدائرة التي مركزها ن فإنها تكون مماسة لقوس شفخ. P39r وجميع الدوائر الباقية التي مركزها على قوس كنم تكون قاطعة لقوس شفخ على نقطتين: أحدهما نقطة ف والأخرى طرف الخط الموازي لخط نف. وهذه الدوائر تكون متصلة متداخلة متراصة من نقطة ش إلى نقطة خ. B94r والتي مركزها ك تقطع قوس شفخ على نقطتي ف ش. والتي مركزها نقطة م تقطع قوس شفخ على نقطتي ف خ. والدوائر الباقية تقطع هذِه القوس على نقط متوالية متراصة. فيكون الجزء L24v من الدائرة التي مركزها ك، الذي يقع في داخل قوسَ شف، أعظم الأجزاء التي تقع في داخل هذه القوس. وكذلك الجزء من الدائرة التي مركزها م، الذي يقع في F122r داخل قوس فخ، أعظم الأجزاء التي تقع في داخل قوس فخ. ويكون الأجزاء من الدوائر الباقية ما قرب منها من الدائرتين المتطرفتين أعظم مما بَعُدَ. BO|

---

452 تكون] تكون FB يكون OL the verb follows the fem sing الدائرة || 453 تكون] تكون FB يكون OL the plural الدوائر requires agreement in fem sing || 454 أحدهما] احدهما BPL احديهما FO || 455 تكون] تكون B يكون OL s.p.F the plural subject الدوائر requires agreement in fem sing || 455 متداخلة] marg B || 457 تقطع] يقطع FOL the verb follows the fem sing الدائرة || 457 على نقطتي ف خ] om B || 458 تقطع] تقطع B يقطع FOL the plural subject الدوائر requires agreement in fem sing. F add من || 459 شف أعظم الأجزاء التي تقع في داخل هذه القوس] marg B || 461 تقع ₂] om L تقع FB يقع O

The circle whose center is K, going through the two points Š F, is tangent to line ŠR, because ŠR [P39v] is perpendicular to ŠK, which is the semi-diameter of this circle. And if this circle, namely, that going through the two points Š F, is tangent to line ŠR, it will take a small part of the sector ŠFR, when the rest of the sector is free. Likewise, the circle that passes through the two points Ḫ F is tangent to line ḪR, and takes a small part of sector ḪFR, when the rest of the sector is free. So do all remaining circles: each and every one of them is tangent to the perpendicular line that comes from [O84r] the endpoint of its diameter to line FT, and takes a small part of the concavity of the arc. |[BPO]

<4.1 Effect of the Size of the Aperture on the Image of the Concave Face>[66]

It has been shown that these circles, [L25r] namely those whose [B94v] centers are on arc KNM, are illuminated with the lights that come from the point of arc ADĞ, which is the base of the concave surface of the cone. Lights that emerge from the arc ADĞ, take two small parts of arc ŠFḪ, and the rest of the concavity of arc ŠFḪ is devoid of light. There is at point F some angle from the intersection of the two circles, whose centers are the two points K M. And because all the equal circles that we have mentioned [P40r] intersect at point F, point F gets a great number of lights. Thus, the angle that is at point F is illuminated by accidental lights that shine from the

66. See Commentary, pp. 135–42.

والدائرة التي مركزها ك التي تمر بنقطتي ش ف هي مماسة لخط شر، لأن شر P39v عمود على شك الذي هو نصف قطر هذه الدائرة. ⌐omL وإذا كانت هذه الدائرة، omL¬ أعني التي تمر بنقطتي ش ف مماسة لخط شر، فهي تأخذ من قطعة شفر جزءًا يسيرًا، ويكون بقية القطعة خالية منها. وكذلك الدائرة التي تمر بنقطتي خ ف تكون مماسة لخط خر فهي تأخذ من قطعة خفر جزءًا يسيرًا، وتكون بقية القطعة خالية منها. وكذلك جميع الدوائر الباقية: كل واحد منها مماسة للعمود الذي يخرج من O84r طرف قطرها على خط فت، فهي تأخذ من تقعير القوس جزءًا يسيرًا. |BPO

وقد تبين أن هذه الدوائر، L25r أعني التي B94v مراكزها على قوس كنم، وهي مضيئة بالأضواء التي تخرج من النقط التي على قوس ادج، الذي هو قاعدة السطح المخروط المقعر. فالأضواء التي تخرج من قوس ادج تأخذ من قوس شفخ جُزءين صغيرين، ويكون بقية تقعير قوس شفخ خالية من الضوء. ويحصل عند نقطة ف زاوية ما من تقاطع الدائرتين، اللتين مركزاهما نقطتا ك م. ولأن جميع الدوائر المتساوية التي ذكرناها P40r تتقاطع

---

464 مركزها ك] ك om F || 465 لأن شز] om L || 465-6 وإذا كانت هذه الدائرة] om L || 468 الدائرة] om F || 468 تكون] om B تكون F يكون OL the verb follows the fem sing الدائرة || 469 وتكون] تكون F ويكون OL the same as above || 473 تخرج] تخرج F يخرج OL s.p.BP the plural الأضواء requires agreement in fem sing || 474 تخرج] تخرج FB يخرج OL s.p.P the same as above || 474 من قوس ادج] repet L || 476-7 ويحصل عند نقطة ف ... نقطتا ك م] marg P || 477 مركزاهما] مراكزهما L || 477 ولأن] لأن P || 477 تتقاطع] مقاطع F يتقاطع OL بتقاطع B the plural الدوائر requires agreement in fem sing

light in the air surrounding them. Some of the light that is inside the concavity of the arc scatters,[67] because light does not come to the perimeter of even one of these circles, except from a single point of arc ADĞ, which is the tip[68] of the light crescent. Yet, the light that comes from a single point is very weak. If the angle at point F is lit, [L25v] and light which is in the concavity of arc ŠFḪ scatters, then the concavity of light becomes a circular concavity, even though this concavity it is not of proper circularity. However, the inequalities in this concavity are not perceived by the sense[69], so that the form of light that appears on the plane parallel to the aperture is a crescent-shaped figure, whose convex arc is of perfect circularity and concave arc is not of perfect circularity, though it looks round to the sense. ||[BO]

<4.2 Effect of the Focal Distance on the Image of the Concave Face>[70]

If afterwards the aperture is moved away from the plane parallel to it, or the plane is moved away from the aperture, the [P40v] distance [B95r] between the centers of the two arcs, which is line FT, will be smaller than the semi-diameter of each one of these

67. Lit.: "Some of the light... split up."
68. Lit.: "the end."
69. See note 25.
70. See Commentary, pp. 142–5.

على نقطة ف، يحصل عند نقطة ف أضواء كثيرة. فتضيء الزاوية التي عند نقطة ف بالأضواء العرضية التي تشرق عليها من الضوء الذي في الهواء المحيط بها. ويتشعَب بعض الضوء الذي في داخل تقعير القوس، لأنه ليس يخرج إلى محيط الدائرة الواحدة من هذه الدوائر ضوء، إلا من نقطة واحدة من قوس ادج، التي هي نهاية الهلال المضيء. والضوء الذي يخرج من نقطة واحدة يكون ضعيفًا جداً. وإذا أصاب الزاوية التي عند نقطة ف L25v وتشعب الضوء الذي في تقعير قوس شفخ، صار تقعير الضوء تقعيرًا مستديرًا، إلا أنه ليس يكون هذا التقعير صحيح الاستدارة. ولكن التفاوت الذي يكون في هذا التقعير لا يحققه الحس، فيكون شكل الضوء الذي يظهر في السطح الموازي للثقب، شكلاً هلاليًا محدبه مستدير صحيح الاستدارة، ومقعره مُستدير ليس بصحيح الاستدارة لكنه عند الحس مستدير. ||BO

ثم إن بوعد الثقب عن السطح الموازي له، أو بوعد السطح عن الثقب، فإن P40v البعد B95r الذي بين مركزي القوسين، الذي هو خط فت، يكون أصغر من نصف قطر كل واحدة من القسي، كما تبين من قبل. فيكون

478 يحصل] يحصل L تحصل B فحصل FO the same as above || 478 فتضيء] فيضي FO || 479 تشرق] تشرق FBL يشرق O the plural subject الأضواء requires agreement in fem sing || 480 بها] om F || 480 ويتشعَب] ويتشعَب FBL ونيشعث O the verb precedes the masc sing الضوء || 483 نقطة 2] om L || 483 أصاب] أصاب B اضات FO اضاءت L || 484 الضوء تقعيرًا] om B || 489 مستدير] مستديرا O || 490 بوعد 1] بوعد FB يوعد OL || 490 بوعد 2] بُوعدَ F بوعد L يوعد BO

arcs, as shown before. Line NF becomes smaller than the semi-diameter of arc KNM and the center of the arc falls within the concavity of arc ŠFḪ. Line FK is greater than line FN, FN is always equal to KŠ, and line FK is greater than line KŠ. Thus, circle ŠṮ cuts line KF under arc ŠF. The same holds for all circles equal to circle ŠṮ whose centers are on arc KNM: The lines that [L26r] come from their centers to point F are greater [F122v] than their semi-diameters. Thus, all these circles have their perimeters below[71] point F. The circle, whose center is point N, passes alone through point F and is tangent to arc ŠFḪ. Each one of the remaining circles cut arc ŠFḪ, and a part of it falls within the concavity of the arc ŠFḪ. However, each one of the parts that fall inside the concavity of the arc ŠFḪ of these circles [P41r] is smaller than that of the circles passing through point F. From that it follows that this concavity is sharper[72] than the first concavity, because the intersection of the light circles is under point F, and the luminous parts that are within arc ŠFḪ are fewer[73] than the lights that are inside the arc of the first concavity. And this implies that whenever the

71. I translate تحت differently: "under arc ŠF" and "below point F."

72. Lit.: "stronger."

73. Lit.: "less."

خط نف أصغر من نصف قطر قوس كنم، ويكون مركز هذِه القوس في داخل تقعير قوس شفخ. فيكون خط فك أعظم من خط فن وفن هو أبدا مثل كش، فخط فك يكون أعظم من خط كش. فدائرة شث تقطع خط كف تحت قوس شف. وكذلك جميع الدوائر المساوية للدائرة شث التي مراكزها على قوس كنم: يكون الخطوط التي L26r تخرج إلى مراكزها من نقطة ف أعظم F122v من أنصاف أقطارها. فجميع هذه الدوائر تكون محيطاتها تحت نقطة ف. وتكون الدائرة التي مركزها نقطة ن فقط تمر بنقطة ف وتماس قوس شفخ. وكل واحدة من الدوائر الباقية تقطع قوس شفخ، ويحصل جزء منها في داخل تقعير قوس ⌐omF شفخ. إلا أن الأجزاء التي تحصل في داخل تقعير قوس ¬omF شفخ من هذه الدوائر كل واحد P41r منها <يكون> أصغر من نظيرِه من الدوائر التي تمر بنقطة ف. فيلزم من ذلك أن يكون هذا التقعير أشد من التقعير الأول، لأن تقاطع الدوائر المضيئة يكون تحت نقطة ف، والأجزاء المضيئة التي في داخل قوس شفخ تكون أقل من الأضواء الذي في داخل هذه القوس من التقعير

---

495 كش]₂ ش L || 495 تقطع] تقطع B يقطع FOL the verb follows the fem sing دائرة || 496 للدائرة] للدائرة B لدائرة FPOL || 497 يكون] يكون OL تكون FB the verb precedes the masc plur الخطوط || 497 تخرج] تخرج FB يخرج OL the plural الخطوط requires agreement in fem sing || 498 post أعظم] من أقطارها فيها scr del F || 498 الدوائر] الدائرة F || 498 تكون] تكون FB يكون OL the plural الدوائر requires agreement in fem sing || 499 وتكون] ويكون OL the verb precedes the fem sing الدائرة || 500 تقطع] يقطع O the verb follows the fem sing كل واحدة || 501 جزء] om FP || 501 قوس شفخ إلا أن الأجزاء التي تحصل في داخل تقعير قوس] 10 words om F || 502 تحصل OL the plural الأجزاء requires agreement in fem sing || 506 تكون] تكون B يكون OL the same as above || 506 في] om B

aperture is far [B95v] from the plane parallel to it, the concavity <subjected> to the light will be increasingly deep. |°

The reverse occurs if the aperture is moved close to the plane parallel to it, or the plane is moved close to it <the aperture>, [L26v] and thus makes the line that is between the centers of the two arcs greater than the semi-diameter of the arc. Therefore line NF is greater than the semi-diameter of arc KNM, line FK is smaller than line FN, and FK is smaller than line KŠ. Consequently, circle ŠṮ cuts the line KF above point F. The same applies to the remaining circles: the luminous parts collected within arc ŠFḪ outnumber[74] the luminous parts collected within [P41v] this arc from the circles passing through point F. From this, [O84v] it follows that the concavity is less <sharp> and smaller <in size>. As a result, whenever the aperture comes closer to the plane that displays its light, the concavity that appears in the light will be less <sharp>. |°

It is clear from all we have explained that, when the light of the Sun at the time of its eclipse, provided it is not a total eclipse, has gone through a narrow aperture and has appeared on a plane parallel to the aperture, then its form is crescent-shaped, its convexity is round of perfect circularity, while its concavity is round to

74. Lit.: "are... more."

الأول. فيلزم من ذلك أن يكون الثقب، كلما بعد B95v عن السطح الموازي له، كان التقعير، الذي في الضوء أكثر وأشد انخماصًا. °ا

ويعرض ضد ذلك إذا قرب الثقب من السطح الموازي له أو قرب السطح منه، وذلك أنه يعرض L26v من ذلك أن يكون الخط الذي بين مركزي القوسين أعظم من نصف قطر القوس. فيكون خط نف أعظم من نصف قطر قوس كنم، فيكون خط فك أصغر من خط فن، فيكون فك أصغر من خط كش. فيكون دائرة شث تقطع خط كف فوق نقطة ف. وكذلك الدوائر الباقية: فيكون الأجزاء المضيئة التي تحصل في داخل قوس شفخ أكثر من الأجزاء المضيئة التي تحصل في داخل P41v هذِه القوس من الدوائر التي تمر بنقطة ف. فيلزم من ذلك O84v أن يكون التقعير أقل وأصغر. فيلزم أن يكون الثقب كلما قرب من السطح الذي يظهر عليه الضوء كان التقعير الذي يظهر في الضوء أقل. °ا

فقد تبين من جميع ما بيناهُ أن ضوء الشمس في وقت كسوفها، إذا لم يستغرق الكسوف جميعها، إذا خرج من ثقب ضيق وظهر على سطح مواز للثقب، فإن شكله يكون هلاليًا، محدبه مستدير صحيح الاستدارة ومقعره

---

507 بعد] بَعُد B بعد FPOL || 511 أعظم من نصف قطر القوس فيكون خط نف] repet F || 513 تقطع] تقطع F يقطع OL the verb follows the fem sing دائرة || 514 تحصل] تحصل FB يحصل OL the plural الأجزاء requires agreement in fem sing || 515 تحصل] تحصل FB يحصل OL the same as above || 516 القوس من] marg P || 517 الثقب] om FO || 519 ضوء] الضوء O || 519 في] om PL || 520 ثقب] الثقب O || 520 مواز] موازي BO || 521 هلاليًا] om F || 521 مستدير صحيح الاستدارة ومقعرة] marg P

the sense only. The farther the aperture from the plane parallel to it, the sharper the concavity of the crescent; [L27r] The closer the aperture to the plane parallel to it, the less sharp the concavity. If [P42r] the remoteness or closeness varies widely, the sharpening or softening of the concavity is perceived by the sense;[75] If the remoteness or closeness varies little, the sharpening or softening of the concavity does not appear. The [B96r] crescent displayed[76] is greater than the ones <that are> similar to the self-luminous crescent. This means that the ratio of (light) to (shade) in its concavity is greater than the ratio of (the light that originates from the Sun) to (the darkness that arises from the concavity).[77] And that is what we wanted to show. |[BOP]||[F]

<4.3 A Condition for Light to Appear Crescent-Shaped or Circular>

It is clear from all that we have explained above that if any circular aperture is facing the luminous part of the Sun, if there is behind the aperture a plane parallel to the plane of the aperture, and if the ratio of (the diameter of the aperture) to (the diameter of the self-luminous body, of which the self-luminous crescent is a part) is no greater than the ratio of (the distance from the aperture to the plane parallel to it) to (the distance between the plane parallel to the aperture and the self-luminous crescent), then the light appears crescent-shaped on the plane parallel to the aperture. ||[O]

[P43r] As a result of what has been found, I say: if [L27v] the ratio of (the distance between the aperture and the plane parallel to the aperture) to (the distance between [B96v] the plane parallel to the aperture and the body) is equal to the ratio of (the

---

75. Lit.: "the increase and decrease." About perception, see note 25.

76. Lit.: "appears."

77. Lit.: "the dark concavity that appears."

مستدير في الحس. فإن الثقب كلما بَعُدَ عن السطح الموازي له كان تقعير الهلال أكثر؛ L27r وكلما قرب الثقب من السطح الموازي له كان التقعير أقل. وإذا كانت P42r زيادة البعد والقرب كثيرة ظهرت الزيادة والنقصان في التقعير للحس؛ وإذا كانت زيادة البعد والقرب يسيرة لم يظهر الزيادة والنقصان في التقعير. فإن هذا B96r الهلال الذي يظهر هو أعظم من الشبيه بالهلال المُضيء، أعني أن نسبة الضوء الذي فيه إلى الظل الذي في تقعيره أعظم من نسبة الضوء الذي يظهر من الشمس إلى التقعير المظلم الذي يظهر منها. وذلك ما أردنا أن نبيّن. F||BOP|

ويتبين من جميع ما بيناه أن كل ثقب مستدير، إذا قوبل به الجزء المضيء من الشمس وكان وراء الثقب سطح مواز لسطح الثقب، وكانت نسبة ⟨قطر الثقب⟩ إلى ⟨قطر الجسم المضيء الذي الهلال المضيء جزء منه⟩ ليست بأعظم من نسبة ⟨بُعد الثقب عن السطح الموازي له⟩ إلى ⟨البعد الذي بين السّطح الموازي للثقب وبين الهلال المُضيء⟩، فإن الضوء يظهر على السّطح الموازي للثقب هلاليًا. O||

P43r وإذ قد تبين ذلك فإنا نقول أنه: إذا L27v كانت نسبة ⟨البعد الذي بين الثقب وبين السطح الموازي للثقب⟩ إلى ⟨البعد الذي بين B96v السطح الموازي للثقب وبين الجسم⟩ كنسبة ⟨نصف قطر الثقب⟩ إلى ⟨عشرة أضعاف

---

522 بَعُدَ] بَعُد B بعد FPOL || 527 بالهلال] om F || 531 مواز] موازي BO || 532 المضيء₂] om L || 533 له] om L || 534 على] عن F || 538 نصف] om P

semi-diameter of the aperture) to (ten times the semi-diameter of the self-luminous body), and if the aperture is circular, then light that is displayed on the plane parallel to the aperture will appear circular, and will not show any concavity. |[PO]||[B]

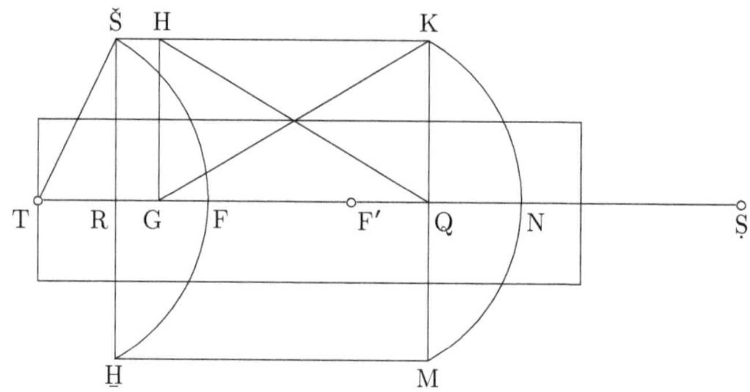

Diagram 4[78]

<4.4 Geometric Demonstration>

<Diagram 4> Let us single out the two arcs KNM ŠFḪ, not to have too many lines <on the figure>. We draw the line NFT and drop ŠRḪ. We draw the two lines ŠK ḪM, and drop the perpendicular KQ. Then QR will be equal to NF, for KŠ is equal to NF. The surface KR is a right-angled parallelogram. If the ratio of (the distance between the aperture and the plane parallel to the aperture) to (the distance between the plane parallel to the aperture and the self-luminous body) is equal to the ratio of (the semi-diameter of the aperture) to (ten times the semi-diameter [F123r] of the self-luminous body), then the line between the centers of the two arcs KNM ŠFḪ is ten times [P43v] the semi-diameter of each of these two arcs. So it is, for it has been

78. Diagram 4 is a (rotated) simplified depiction of the two crescents KLNM, ŠYḪF, to determine the length of segment RG, i.e., the depth of the cavity in the rounded image.

نصف قطر الجسم المضيء⟩، وكان الثقب مستديرًا، فإن الضوء الذي يظهر على ذلك السطح الموازي للثقب يكون مستديرًا، ولا يكون فيه شيء من التقعير. |B||PO|

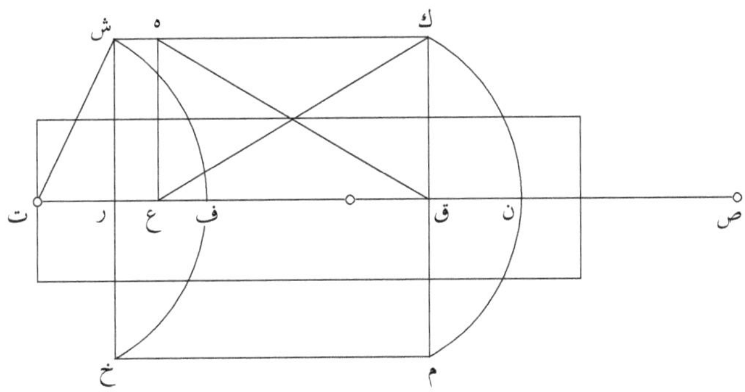

< شكل ٤ >

<Fig.4> فلنفرد قوسي كنم شفخ حتّى لا تكثر الخطوط. ونخرج خط نفت ونصل شرخ. ونصل خطي شك خم ونخرج عمود كق. فيكون قر مساويا لنف، لأن كش مساوي لنف. ويكون سطح كر متوازي الأضلاع قائم الزوايا. وإذا كانت نسبة ⟨البعد الذي بين الثقب وبين السطح الموازي للثقب⟩ إلى ⟨البعد الذي بين السطح الموازي للثقب وبين الجسم المضيء⟩ كنسبة ⟨نصف قطر الثقب⟩ إلى ⟨عشرة أضعاف نصف قطر F123r الجسم المضيء⟩ كان الخط الذي بين مركزي قوسي كنم شفخ عشرة

543 فلنفرد] فلنفرد FPOL فلبفرد B || 543 تكثر] تكثر F يكثر OL s.p. BP the plural الخطوط requires agreement in fem sing || 549 مركزي] المركزي L || 547 الذي] om F || 545 مساويا] مساويا POL مساوبا F مساو B || 545 قر] قنه F ||

proven <Diagram 1B> that (the line between the two centers)[79] to (the line <ṢX>, [L28r] which is the extension[80] of line BṢ up to the line coming from point T to the center of the aperture, between the endpoint of the line mentioned and point Ṣ)[81] is equal to the ratio of (the distance between the plane parallel to the aperture and the aperture) to (the distance between the aperture and the self-luminous body). [O85r] If the ratio of (the distance between the aperture and the plane parallel to the aperture) to (the distance between the plane parallel to the aperture and the self-luminous body) is equal to the ratio of (the semi-diameter of the aperture) to (ten times the semi-diameter of the self-luminous body), then (the line <ṢX>, which is the extension of line BṢ between the endpoint of the line coming from point T to the center of the aperture that ends on the self-luminous body, and point [B97r] Ṣ) is (ten times the line ṢD).[82] Therefore, the ratio of (the line between the two centers)[83] to (ten times the semi-diameter of the self-luminous body) [P44r] is equal to the ratio of (the line between the centers of the two arcs KNM ŠFḪ) to (ten times the semi-diameter of each of these two arcs). And so it has been established that (the line between the two centers) to (the line <ṢX>, which is the extension of line BṢ up to the line coming from point T to the center of the aperture, between the endpoint of

79. This is the distance between the crescents $TF' = NF$.

80. Lit.: "separates."

81. This awkward phrasing is because line ṢX is not drawn. I have reworded it.

82. On this hard-to-read passage, see Commentary, pp. 117 sq.

83. "The line between the two centers" refers in general to distance TF′. If this were the case here, we should read: $TF'/10\ ṢD = TF'/10\ NF' = 1$, which makes no sense. Thus, I read "the line between the two points" <Ṣ and X>.

أضعاف P43v نصف قطر كل واحد من هذين القوسين. وذلك أنه قد تبين أن ⟨الخط الذي بين المركزين⟩ نسبته إلى ⟨الخط الذي L28r يفصله الخط الذي يخرج من نقطة ت إلى مركز الثقب، من خط ب‍ص وما يتصل به، الذي بين طرف الخط الذي ذكرناه وبين نقطة ص⟩ كنسبة ⟨البعد الذي بين السطح الموازي للثقب وبين الثقب⟩ إلى ⟨البعد الذي بين الثقب وبين الجسم المضيء⟩. O85r وإذا كانت نسبة ⟨البعد الذي بين الثقب وبين السطح الموازي للثقب⟩ إلى ⟨البعد الذي بين السطح الموازي للثقب وبين الجسم المضيء⟩ كنسبة ⟨نصف قطر الثقب⟩ إلى ⟨عشرة أضعاف نصف قطر الجسم المضيء⟩، فالخط ⟨الذي ينفصل من خط ب‍ص⟩ وما يتصل به الذي بين طرف الخط الذي يخرج من نقطة ت إلى مركز الثقب وينتهي إلى الجسم المضيء وبين نقطة B97r ص⟩ يكون ⟨عشرة أضعاف خط ص‍د⟩. فيكون نسبة ⟨الخط الذي بين المركزين⟩ إلى ⟨عشرة أضعاف نصف قطر الجسم المضيء⟩ P44r ⌐omFBOL كنسبة ⟨الخط الذي بين مركزي قوسي ك‍ن‍م ش‍ف‍خ⟩ إلى ⟨عشرة أضعاف نصف قطر كل واحد من هذين القوسين⟩. وذلك أنه قد تبين أن ⟨الخط الذي بين المركزين⟩ نسبته إلى ⟨الخط الذي يفصله الخط الذي يخرج من نقطة ت إلى مركز الثقب، من خط ب‍ص وما يتصل به، الذي

---

550 أنه] om L || 551 بين المركزين نسبته إلى الخط] marg B || 555 وإذا كانت نسبة ... الجسم المضيء] marg B || 562 كنسبة ⟨الخط الذي ... ذكرناه وبين نقطة ص⟩] 56 words om FBOL by homoioarkhton كنسبة or by homoioteleuton وبين نقطة ص⟩ || 562 كنسبة] P lectio incerta

the line mentioned, to point Ṣ) is equal to the ratio of (the distance between the plane parallel to the aperture and the aperture) to (the distance between that aperture and the self-luminous body). |P||O

However, as the ratio of (the distance between the aperture and the plane parallel to the aperture) to (the distance between the plane parallel to the aperture and the self-luminous body) is equal to the ratio of (the semi-diameter of the aperture) to (ten times the semi-diameter of the self-luminous body), and (the line <ṢX>, which is the extension of line BṢ, between the endpoint of the line coming from point T to the center of the aperture that ends on the self-luminous body, and point Ṣ) is (ten times the line ṢD), then the ratio of (the line between the two centers) to (ten times the semi-diameter of the self-luminous body) is equal to the ratio of [P44v] (the distance between <the aperture and> the plane parallel to the aperture) to (the distance between the aperture and the self-luminous body). |P||O

Also, it is established that the ratio of (the semi-diameter of arc KNM) to (the semi-diameter of the self-luminous body) is equal to the ratio of (the distance between the plane parallel to the aperture and the aperture) to (the distance between the aperture and the self-luminous body). [L28v] This implies that the ratio of (the line between the two centers)[84] to (ten times the semi-diameter of the self-luminous body)

84. This is again the distance between the crescents TF′ = NF.

بين طرف الخط الذي ذكرناه وبين نقطة ص) ⌝omFBOL كنسبة (البعد الذي بين السطح الموازي للثقب وبين الثقب) إلى (البعد الذي بين الثقب وبين الجسم المضيء). |P||O|

omFBOL⌝ وإذا كانت نسبة (البعد الذي بين الثقب وبين السطح الموازي للثقب) إلى (البعد الذي بين السطح الموازي للثقب وبين الجسم المضيء) كنسبة (نصف قطر الثقب) إلى (عشرة أضعاف نصف قطر الجسم المضيء) (فالخط الذي ينفصل من خط ب‌ص وما يتصل به، الذي بين طرف الخط يخرج من نقطة ت إلى مركز الثقب وينتهي إلى الجسم المضيء وبين نقطة ص) يكون (عشرة أضعاف خط ص‌د). فيكون نسبة (الخط الذي بين المركزين) إلى (عشرة أضعاف نصف قطر الجسم المضيء) كنسبة P44v (البعد الذي بين <الثقب وبين> السطح الموازي للثقب) إلى (البعد الذي بين الثقب وبين الجسم المضيء). ⌝omFBOL |P||O|

وقد تبيّن أيضا أن نسبة (نصف قطر قوس ك‌ن‌م) إلى (نصف قطر الجسم المضيء) كنسبة (البعد الذي بين السطح الموازي للثقب وبين الثقب) إلى (البعد الذي بين الثقب وبين الجسم المضيء). L28v فيلزم من ذلك أن يكون نسبة (الخط الذي بين المركزين) إلى (عشرة أضعاف نصف قطر الجسم المضيء) كنسبة (نصف قطر قوس ك‌ن‌م) إلى (نصف قطر الجسم المضيء).

---

566 كنسبة] resumption of the text common to FBPOL || 569 وإذا كانت نسبة ... السطح الموازي] 83 words om FBOL by homoioteleuton الثقب إلى البعد الذي بين الثقب وبين الجسم المضيء || 578 وقد] resumption of the common text

is equal to the ratio of (the semi-diameter of arc KNM) to (the semi-diameter of the self-luminous body). Alternately,[85] the ratio of (ten times the semi-diameter of the self-luminous body) to (the semi-diameter of the self-luminous body) is equal to the ratio of (the line between the two centers) to (the semi-diameter of arc KNM). Besides, if the ratio of (the distance between the aperture and the plane parallel to the aperture) to (the distance between the plane parallel to the aperture and the self-luminous body) is equal to the ratio of (the semi-diameter of the aperture) to (ten times the semi-diameter of the self-luminous body), then line KŠ, equal to line NF, is equal to ten times the line that is between the two centers, [P45r] which is line FT. The surface KR is a rectangle. Thus, the line that connects the two points K R, which is the diameter of surface KR, is greater than KŠ. Assume that KG is equal to KŠ. Point G will be between the two points F R. The square of KG is equal to the square of QR. We draw QR on the side of Q towards Ṣ, and make QṢ equal to QR. Therefore, [L29r] ṢG multiplied by GR plus the square of QG [B97v] is equal to the square of QR. Then ṢG multiplied by GR plus the square of QG is equal to the square of KG. But the square of KG is equal to the square of KQ plus the square of QG. Then ṢG multiplied by GR plus the square of QG is equal to the square of KG, and the square of KG is equal to the square of KQ plus the square of QG. It results that the square of KQ is equal to ṢG multiplied by GR, and the square of KQ is equal to the square of ŠR. Therefore, ṢG multiplied by GR is equal to the square of ŠR. Arc ŠF is less

---

85. *Alternando* (*Elements* V, def. 12).

فبالتبديل يكون نسبة ⸢عشرة أضعاف نصف قطر الجسم المضيء⸣ إلى ⸢نصف قطر الجسم المضيء⸣ كنسبة ⸢الخط الذي بين المركزين⸣ إلى ⸢نصف قطر قوس كنم⸣. فإذا كانت نسبة ⸢البعد الذي بين الثقب وبين السطح الموازي للثقب⸣ إلى ⸢البعد الذي بين السطح الموازي للثقب وبين الجسم المضيء⸣ كنسبة ⸢نصف قطر الثقب⸣ إلى ⸢عشرة أضعاف ⌝omB نصف قطر الجسم المضيء⸣، فإن خط كش، المساوي لخط نف، المساوي للخط الذي بين المركزين، يكون عشرة أضعاف ⌜omB P45r خط فت. وسطح كر قائم الزوايا. فالخط الذي يتصل بين نقطتي ك ر، الذي هو قطر سطح كر، هو أعظم من كش. فتجعل كع مثل كش. فيكون نقطة ع فيما بين نقطتي ف ر. ويكون مربع كع مثل مربع قر. ونخرج قر في جهته ق إلى ص، ونجعل قص مثل قر. فيكون L29r ضرب صع في عر مع مربع قع B97v مثل مربع قر. فضرب صع ⌝omP في عر مع مربع قع مثل مربع كع. لكن مربع كع مثل مربع كق مع مربع قع. ⌝omBL فمربع صع ⌜omP في عر مع مربع قع مثل مربع كع. لكن مربع كع مثل مربع كق مع مربع قع ⌜omBL. فمربع كق مثل ضرب صع في عر، ومربع كق مثل مربع شر. فضرب صع في عر مثل مربع شر. وقوس شف هي أقل من ربع دائرة،

---

583 فبالتبديل] فبالأبدال L || 587-8 نصف قطر الجسم ... يكون عشرة أضعاف] 18 words om B || 590 بين] om B || 591 من] om L || 591 ante بين] وما add B || 594 فمربع صع ... في عر مع مربع قع] 18 words om P || 595-6 فمربع صع في ... كق مع مربع قع] 19 words om BL || 595 صع] صدع F || 598 شر] شب O

than a quarter of a circle, and each crescent is bounded by the arcs of two equal circles; thus the concave arc is less than half of the circle, because whatever two equal circles intersecting <with each other>, the line that connects their intersections is a chord in each of them. It is thus smaller than their <common> diameter.[86] [P45v] Line ŠḪ is smaller than the diameter of arc ŠFḪ. Arc ŠFḪ is less than half of the circle, line ŠR is smaller than line FT, and line ṢR is twenty times line FT. Thus line ŠR is <less than> half tenth of line ṢR, and less than one tenth of line QR. Therefore, the square of ŠR is less than one hundredth part of the square of QR. And ṢG multiplied by GR is less than one hundredth part of the square of QR. We raise [L29v] at point G the vertical line GH. We draw QH, which is equal to line KG, and equal to line QR. Point H is thus on the perimeter of the circle whose center is Q and the semi-diameter is QR. Thus, the ratio RG to GH is equal to the ratio HG to GṢ. And the ratio TG to GṢ is equal to the ratio HG to GṢ doubled. HG is half tenth of GṢ, rounded off. Thus line RG is the four hundredth part of line [F123v] RṢ, rounded off. To the extent that it is the four hundredth part of line RṢ, <GR> is the twentieth part of line FT and the fortieth part of the diameter [B98r] of circle ŠFḪ. Thus, line GR is the fortieth part of the diameter of arc [P46r] ŠFḪ. The circle whose center is K passes through the two points Š G. And this circle has been shown to be the light circle, whose light originates from the point of the self-luminous crescent, which is point Ğ. |[FP]||[B]

86. Lit.: "the diameter of each of them."

وذلك أن كل هلال يحيط به قوسان من دائرتين متساويتين، فإن القوس المقعرة منهما تكون أقل من نصف دائرة، لأن كل دائرتين متساويتين يتقاطعان، فإن الخط الذي يتصل بين تقاطعهما هو وتر في كل واحد منهما. فهو أصغر من قطر كل واحدة منهما. P45v فخط شخ أصغر من قطر قوس شفخ. وقوس شفخ أقل من نصف دائرة. فخط شر أصغر من خط فت، وخط صر هو عشرون ضعفا لخط فت. فخط شر <أقل من> نصف عشر خط صر، وأقل من عشر خط قر. فمربع شر أقل من مائة جُزء من مربع قر. فضرب صع في عر أقل من جُزء من مائة جُزء من مربع قر. ونخرج L29v من نقطة ع عمود عه. ونصل قه، فيكون مساويا لخط كع فهو مساو لخط قر. فيكون نقطة ه على محيط الدائرة التي مركزها ق ونصف قطرها قر. فيكون نسبة رع إلى عه كنسبة هع إلى عص، فنسبة تع إلى عص كنسبة هع إلى عص مثناه. وهع نصف عشر عص على التقريب. فخط رع جُزء من أربع مائة جُزء من خط F123v رص على التقريب. وبالمقدار الذي به خط رص أربع مائة جُزء، به خط فت عشرون جُزءًا، وبه قطر B98r دائرة شفخ أربعين جُزءًا. فخط عر جُزء من

603 قوس شفخ] om B || 604 <أقل من>] as ŠR is smaller than FT, ŠR is necessarily smaller than one twentieth of ṢR, hence أقل من || 608 مساو] مساوي OB || 609 قر] قر FBPL قت O || 609 رع] رع FBPL تع O || 610 تع] رع FBPL تع O. Neither lettering makes sense, thus I opt for تع. This is a lectio incerta for letters ر and ت are interchanged throughout the passage, lines 609–613 || 611 رع] رع FBPL تع O || 611 رص] رص FBPL تص O || 613 post وبه] نقطتين scr del قطر corr F || 613 عر] عر FBPL عت O || 613 جزء من] om O

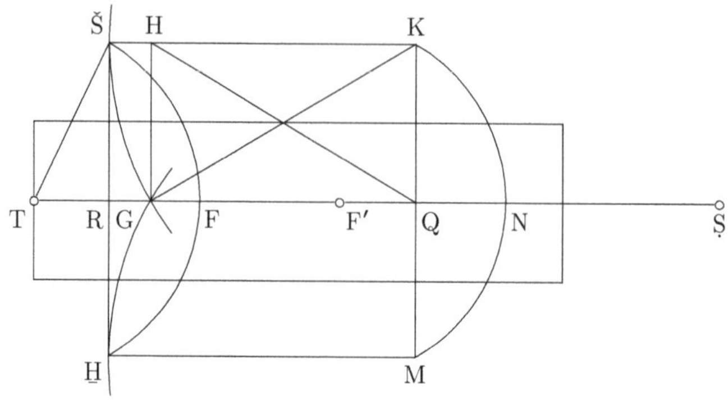

Diagram 4B

<Diagram 4B> Arc ŠFḪ itself is caused by the <light> cone, whose base is the concave arc of the light crescent and the apex is point Ḥ. All light circles, whose centers are on arc [O85v] KNM, cut the arc ŠFḪ. A part of each of them falls inside the arc [L30r] ŠFḪ. These circles are interlocked and overlap one another. The circle of center K is tangent to line ŠR at point Š, and ends at point G. The circle of center M is tangent to line ḪR at point Ḫ, and ends at point G on line FR. The full concavity, which is set up by arc ŠFR, is the sector bound by the arc ŠFḪ and line ŠḪ. If this sector is filled with light, the concavity will disappear in full. If the two circles whose centers are the two points K M [P46v] are illuminated, then they will fill the concavity of the sector with light, for these two circles pass through the points Š G Ḫ and are tangent to line ŠḪ, and the remaining light circles under these two circles are interlocked with them. Nothing remains from the concavity of arc ŠFḪ, except a small

أربعين جُزءًا من قطر قوس P46r شف خ. والدائرة التي مركزها ك تمر بنقطتي ش ع. وهذه الدائرة قد تبين أنها هي الدائرة المضيئة التي يخرج إليها الضوء من نقطة من الهلال المضيء التي هي نقطة ج. |[FP]||[B]|

وقوس شف خ هي التي يحدثها المخروط الذي قاعدته القوس المقعرة من الهلال المضيء ورأسه نقطة ح. وجميع الدوائر المضيئة التي مراكزها على قوس O85v ك ن م تقطع قوس شف خ. ويحصل من كل واحد منها جُزء في داخل قوس L30r شف خ. ويكون هذه الدوائر متصلة متداخلة. والدائرة التي مركزها ك تكون مماسة لخط شر على نقطة ش، وتنتهي إلى نقطة ع. والدائرة التي مركزها م تكون مماسة لخط خر على نقطة خ، وتنتهي إلى نقطة ع من خط فر. وجميع التقعير الذي يوجبه قوس شف ر هو القطعة التي يحيط بها قوس شف خ وخط ش خ. فإذا امتلأت هذه القطعة بالضوء، بطل التقعير بالجملة. والدائرتان اللتان مركزاهما نقطتا ك م، P46v إذا كانتا مضيئتين، فهما يملآن تقعير القطعة ضوءاً، لأن هاتين الدائرتين تمران بنقط ش ع خ وتماسان خط ش خ والدوائر الباقية المضيئة تحت هذه الدائرتين ومتصلات بهما. فليس يبقى من تقعير قوس شف خ، إلاّ جُزء

615 يخرج] تخرج FOL the verb precedes the masc sing الضوء || 615-6 إليها الضوء من نقطة] الضوء اليها من نقطة transp L || 616 من[2]] om FP || 619 تقطع] يقطع FOL the plural subject الدوائر requires fem sing || 621 تكون] يكون F تكون OL the verb تكون follows the fem sing الدائرة || 621 وتنتهي] وينتهي FOL || 622 تكون] يكون FB تكون OL the verb تكون follows the fem sing الدائرة || 622 وتنتهي] وينتهي FOL the same as above || 624 التي] om L || 624 امتلأت] امتلت FBO || 625 بالجملة] om F || 625 نقطتا] نقطتان scr del نقطتا corr B || 626 يملآن] يمليان P يملآن BOL تمليان F || 627 وتماسان] وتماسان L ويماسان BO s.p.FP the plural الدوائر requires fem sing || 628 إلاّ جُزء] repet P

part that is not perceived when the sense falls[87] on point G. All the remaining light circles cut the line [B98v] FR near point G. Point G receives many lights from which accidental lights illuminate line GR. Line GR is very small, thus making the shadow disappear, which is very small at point G. But if the shadow [L30v] which is at point G is hidden, the perimeter of light which is inside the concavity of arc ŠFḪ becomes circular. Thus, the perimeter of the whole light appearing on the plane parallel to the aperture, which is the convex perimeter, turns out to be circular. The curve of the convex arc overlaps with the arc that passes through the two points Š G, because this arc is a part of the circle with center K, whose perimeter is a part of the [P47r] convex arc. All light that appears on the plane parallel to the aperture begins to appear circular. If the ratio of distance to distance, which we have mentioned, is equal to the ratio of the diameter of the aperture to an amount greater than ten times the semidiameter of the self-luminous body, then light becomes rounder, for line KŠ increases, the circle that passes through the two points Š G increases, and line GR diminishes. Therefore, the roundness is sound. |[P]||[BO]

We have proved, amongst others, that if the ratio of (the distance between the aperture and the plane parallel to the aperture) to (the distance between the plane

87. Lit.: "when the sense is at point G."

يسير ليس له قدر عند الحس وهو عند نقطة ع. وجميع الدوائر الباقية المضيئة تقطع خط B98v فر، وتكون قريبة من نقطة ع. فيحصل عند نقطة ع أضواء كثيرة، فيشرق منها أضواء عرضية على خط عر. وخط عر هي غاية الصغر، فيختفي الظل الذي في غاية الصغر الذي عند نقطة ع. وإذا خفي الظل L30v الذي عند نقطة ع، صار محيط الضوء الذي في داخل تقعير قوس شفخ مستديراً. فمحيط جميع الضوء الذي يظهر في السطح الموازي للثقب، الذي هو المحيط المحدب، قد تبيّن أنه مستدير. واستدارة القوس المحدبة متصلة بالقوس التي تمر بنقطتي ش ع، لأن هذه القوس هي من محيط الدائرة التي مركزها ك التي محيطها جزء من القوس المحدبة. P47r فيصير جميع الضوء الذي يظهر في السطح الموازي للثقب يظهر مستديرًا. وإذا كانت نسبة البعد إلى البعد، اللذين ذكرناهما، كنسبة قطر الثقب إلى مقدار أعظم من عشرة أضعاف نصف قطر الجسم المضيء، كان الضوء أشد استدارة، لأن خط كش يكون أعظم فيكون الدائرة التي تمر بنقطتي ش ع أعظم، فيكون خط عر أصغر. فتكون الاستدارة أصحّ. |P||BO|

630 تقطع] تقطع B يقطع FO بقطع L the plural subject الدوائر requires agreement in fem sing || 630 وتكون] وتكون B ويكون FOL the verb precedes the fem sing نقطة || 631 وخط عر] marg F || 636 ش ع] ش ت BO ش ث FPL. In Fig.3, point ث is on line كش, with ثك = كش. As it does not appear in Fig.4, I replace it by point ع for the sake of clarity || 637 التي] التي FOL إلى BP || 639 إلى البعد] om B || 641 خط] repet L || 642 خط عر أصغر فتكون] repet O

parallel to the aperture and the self-luminous body) is equal to the ratio of (the semi-diameter of aperture) to (an amount no smaller than ten times the semi-diameter [L31r] of the self-luminous body), then light that appears on the plane parallel to the aperture will be circular. And this is what we wanted to show. |[PO]||[FB]

<5. Analysis of the Image in the Case of the Moon>[88]

<5.1 Conditions for the Image of the Moon to be Crescent-Shaped>

[B99r] From what we have demonstrated, it is clear that the light of the Moon, at the time of the eclipse or at the time of being a crescent, passes through a circular aperture and appears on a plane parallel to the aperture. If the aperture is one of [P47v] the apertures similar to the aperture faced by the Sun, and by which its light goes through at the time of the eclipse, and if it appears on a plane parallel to the aperture, then the light <of the Moon> that appears on that plane should be crescent-shaped. <And yet,> the light of the Moon, which emerges from the aperture and appears on the plane behind the aperture at the same distance, appears circular and does not show any concavity whatsoever.

This fits with what Ptolemy[89] has found in his book entitled the *Almagest*, namely that the diameter of the Sun is eighteen times and four-fifths as the diameter of the Moon. The ratio of (the diameter of the Sun) to (the diameter of the Moon) is

---

88. See Commentary, pp. 148–58.

89. This is a reference to Ptolemy, *Almagest* V, 16: "ἡ μὲν τῆς γῆς ἄρα διάμετρος τῆς σεληνιακῆς τριπλασίων ἐστιν καὶ ἔτι τοῖς δυσὶ πέμπτοις μείζων, ἡ δὲ τοῦ ἡλίου τῆς μὲν σεληνιακῆς ὀκτοκαιδεκαπλασίων καὶ ἔτι τοῖς δ̲ πέμπτοις μείζων τῆς δὲ γῆς πενταπλασίων καὶ ἔτι τῷ ἡμίσει ἔγιστα μείζων," which corresponds to "Therefore the earth's diameter is $3\frac{2}{5}$ times the moon's and the sun's diameter is $18\frac{4}{5}$ times the moon's and $5\frac{1}{2}$ times the earth's" (Toomer 1984: 257; see Halma 1813: 347). Sabra discussed this reference in this way: "Finally, and on the basis of faulty measurements derived from the *Almagest* (the solar diameter is eighteen and four-fifths times the lunar diameter), Ibn al-Haytham tries to give an answer to the question posed at the beginning of his treatise. His answer recognizes the theoretical possibility of obtaining a crescent image for the crescent Moon, but he considers that the conditions of the arrangement are such that such an image will be too faint to be visible" (Sabra 1989, I: 1).

فقد تبين مماّ بيناهُ أنه إذا كانت نسبة ⸢البعد الذي بين الثقب وبين السطح الموازي للثقب⸣ إلى ⸢البعد الذي بين السطح الموازي للثقب وبين الجسم المضيء⸣ كنسبة ⸢نصف قطر الثقب⸣ إلى ⸢مقدار ليس بأصغر من عشرة أضعاف نصف قطر L31r الجسم المضيء⸣، فإن الضوء الذي يظهر على السطح الموازي للثقب يكون مستديرًا. وذلك ما أردنا أن نبيّن. FB||PO|

B99r وإذ قد تبين ذلك فقد تبيّن أن القمر، في وقت كسوفه، وفي وقت كونه هلالا، إذا خرج ضوءه من ثقب مستدير وظهر على سطح مواز للثقب، وكان الثقب من P47v الثقوب التي إذا قوبل بمثلها الشمس في وقت كسوفها، وخرج ضوءها من ذلك الثقب، وظهر على سطح مواز للثقب، كان الضوء الذي يظهر على ذلك السطح هلاليًا. فإن ضوء القمر، الذي يخرج من ذلك الثقب ويظهر على سطح بُعده من الثقب مثل ذلك البعد، يظهر مستديرًا ولا يظهر فيه شيء من التقعير.

وذلك أن بطلميوس قد بيّن في كتابه الموسوم بالمجسطي، أن قطر الشمس مثل قطر القمر ثمانية عشر مرة وأربعة أخماس مرة. ونسبة ⸢قطر الشمس⸣ إلى ⸢قطر القمر⸣ كنسبة ⸢بُعد الشمس من الأرض⸣ إلى ⸢بُعد القمر من الأرض⸣، لأن الشمس والقمر يحيط بهما مخروط واحد رأسه مركز البصر

650 ضوءه] ضوء L || 650 مواز] موازي BO || 651 من] من FBPL مثل O || 653 ضوء] الضوء O || 656 الموسوم] المعروف L || 656 أن$_2$] إلى L || 657 ثمانية] ثمنية FBOL

equal to the ratio of (the distance from the Sun to the Earth) to (the distance from the Moon to the Earth), because the Sun and the Moon are enclosed in one cone, whose apex is the center of vision which is on the surface of the Earth. Thus the distance from the Sun to the Earth [L31v] is eighteen times and four-fifths of the distance from the Moon to the Earth. Therefore, if the Moon is facing an aperture whose diameter is one part out of eighteen and four-fifths[90] of the diameter of the aperture opposite to the Sun, if the sunlight which is made crescent-shaped passes through it, and if the distance of the plane that shows the moonlight to the aperture is one part out of eighteen[91] and four-fifths of [B99v] the distance from the plane that shows the sunlight to the aperture through which passes [O86r] the sunlight, the ratio of (the distance from the aperture to the plane parallel to the aperture) to (the distance from the plane to the Moon) will be as (the diameter of the aperture) to (the diameter of the Moon). These are the conditions for the moonlight to appear crescent-shaped. |[BO]

<5.2 Material Impossibility for These Conditions to be Fulfilled>

If the diameter of the aperture is one part out of eighteen of the diameter of the aperture by which the sun<light> is tested,[92] then the full area of the aperture by which the moon<light> is tested should be one part out of three hundred and twenty-four of the full area of the aperture by which the sun<light> is tested. If the diameter of the aperture by which the sun<light> is tested is as a grain of barley,[93] then one part out of three hundred and twenty-four [L32r] of the full area[94] of this aper-

---

90. F ends at time 0.922.

91. P ends at time 0.925.

92. Each of the following occurrences can be read "tested" or "experimented."

93. On the grain of barley as the minimum size of the aperture, see Commentary, pp. 102–3.

94. Lit.: "surface."

الذي هو على وجه الأرض. فبُعد الشمس من الأرض L31v مثل بُعد القمر من الأرض ثمانية عشر مرة وأربعة أخماس. فإذا قوبل القمر بثقب قطره جزء من ثمانية عشر جُزءًا وأربعة أخماس ⌐omF من قطر الثقب الذي قوبل به الشمس، وظهر ضوء الشمس الذي ينفذ فيه هلاليًا، وكان بُعد السطح الذي يظهر عليه ضوء القمر من الثقب جُزءًا من ثمانية عشر ⌐omP جُزءًا وأربعة أخماس B99v من البعد الذي بين السطح، الذي يظهر عليه ضوء الشمس، وبين الثقب الذي يخرج منه ضوء O86r الشمس، كانت نسبة ⟨بُعد الثقب عن السطح الموازي للثقب⟩ إلى ⟨بُعد السطح عن القمر⟩ كنسبة ⟨قطر الثقب⟩ إلى ⟨قطر القمر⟩. فعند ذلك يجب أن يظهر ضوء القمر على السّطح هلاليًا. |BO

وإذا كان قطر الثقب جُزءًا من ثمانية عشر جزءًا من قطر الثقب الذي يعتبر به الشمس، فإن جميع سطح الثقب الذي يعتبر به القمر، يجب أن يكون جُزءًا من ثلثمائة وأربعة وعشرين جُزءًا من جميع سطح الثقب الذي يعتبر به الشمس. فإذا كان قطر الثقب الذي يعتبر به الشمس عرض شعيرة واحدة، كان الجزء من ثلثمائة وأربعة وعشرين L32r جُزءًا من جميع سطح هذا الثقب جُزءًا غير محسوس، لأنه يكون بمنزلة النقطة. ويكون الضوء الذي يخرج

---

660 هو] om P || 661 ثمانية] ثمنية FBOL || 661 عشر] عشر BPL عشرة FO || 662 من₁] ومن BL || 662 ثمانية] ثمنية FBOL || 662 ante من₂] F ends at time 0.922 || 664 ثمانية] ثمنية BOL || 664 post عشر] P ends at time 0.925 || 668 يجب] يجب O نجب L the verb follows the masc sing ذلك || 670 ثمانية] ثمنية BOL || 673 ante شعيرة] ضوء scr del O || 674 ثلثمائة] ثلثمايه BOL

ture is an imperceptible size, because it amounts to a point. Light that comes from it is imperceptible, especially the light of the Moon, which is faint. Once it has come out from such point-shaped aperture, the light that is received on the plane opposite to it turns out to be imperceptible, because of its smallness and for being hidden. Therefore, if the moon hits an aperture of this nature at the time of being crescent-shaped, it will not appear crescent-shaped because it is hidden. |°

The diameter of the aperture by which the light of the Sun is tested is eighteen and four-fifths times the diameter of the aperture that would allow the light of the Moon to appear crescent-shaped. It has been shown with demonstration that if any self-luminous body was facing an aperture with a diameter not smaller than ten times the diameter of the aperture showing its crescent-shaped light, then the light that came out from it [B100r] to the plane showing the light crescent-shaped, was circular. If the Moon is facing an aperture like that by which the sunlight appears crescent-shaped, and if the moonlight passes through it and attains the plane whose distance from the aperture is one part out of eighteen times the distance showing the sunlight crescent-shaped, then the light will appear circular,[95] [L32v] as shown before by proof. If the moonlight being processed through a suitable aperture is received on a plane close to the aperture, it should appear circular. Then, if the aperture moves away from the plane on which the light is displayed, the light weakens, and what weakens first is its edges. |°

95. The proof contains an error, see Commentary, pp. 149–50.

منه غير محسوس، وخاصة ضوء القمر فإن ضوء القمر ضعيف. وإذا خرج من ثقب بمنزلة النقطة، كان الضوء الذي يحصل على السطح المقابل له غير محسوس لصغره ولِخَفائه. فالثقب الذي بهذه الصفة، إذا قوبل به القمر في حال كونه هلالا، لم يظهر هلاليًا لخفائه. ا°

والثقب الذي يعتبر به ضوء الشمس هو الذي قطرهُ مثل قطر الثقب الذي كان يجب أن يظهر ضوء القمر منه هلاليًا، ثمانية عشر مرة وأربعة أخماس. وقد تبين بالبرهان أن كل جسمٍ مُضيء، إذا قوبل بثقب قطره ليس بأصغر من عشرة أضعاف قطر الثقب الذي يظهر ضوءه هلاليًا، فإن الضوء الذي يخرج منه، B100r إذا حصل على السطح الذي كان يظهر عليه الضوء هلاليًا، ظهر مستديرًا. فإذا قوبل القمر بثقب من مثله يظهر ضوء الشمس هلاليًا، ونفذ فيه ضوء القمر وحصل على السطح الذي بُعده من الثقب جُزء من ثمانية عشر جُزءًا من البعد الذي يظهر عنده ضوء الشمس هلاليًا، وجب أن يظهر الضوء مستديرًا L32v كما تبين من قبل بالبرهان. فضوء القمر الذي ينفذ من ثقب مقتدر إذا حصَل على سطح قريب من الثقب، يجب أن يظهر مستديرًا. ثم إذا بوعِد الثقب عن السطح الذي يظهر عليه الضوء، ضعف الضوء وأول ما يضعف منه حواشيه. ا°

---

678 ولِخَفائه] وخفايه BL وخفاه O || 679 لخفائه] لخفايه BOL || 681 ثمانية] ثمنية BOL || 681 عشر] عشر BO عشرة L || 684 يظهر عليه الضوء] الضوء يظهر عليه transp L || 685 مثله] ميله O || 687 ثمانية] ثمنية BO || 687 ثمانية عشر جزءًا] om L || 688 كما تبين] كما نبين O لما نبين L لما ىس B || 690 بوعِد] يوعد BOL

Assuming that the light coming from the light of the Moon was crescent-shaped, its concavity would disappear with the remoteness of the aperture, because the parts of the light that weaken first with the remoteness are the angles and edges.[96] If what had made the moonlight appear crescent-shaped at short distance moved away from the aperture, then it is its angles that would vanish first,[97] then its edges, so that it would become circular. Thus if <the image> were circular while being close to the aperture, it would be <even> more circular with the remoteness. ||°

<As for> the distance that meets the ratio required for the moonlight passing through a suitable[98] aperture to appear crescent-shaped, if the moonlight arrives with that distance, it will fade out and disappear, for every light coming from an aperture vanishes when the aperture is moved away at a great distance. ||°

The same holds true with the light of the Sun, if it comes from a wide, circular aperture and stands at a distance whose ratio requires that its light appears crescent-shaped. However, that distance may vary and perhaps one cannot [B100v] find on the face of the Earth any place whose distance to the aperture meets the required ratio. And if [L33r] a place of this kind were found, the eyesight of the experimenter who is at the aperture would not perceive what stands in that position, because of the discrepancy of the distances. However, if the light of the Sun that came out from the wide

96. See Commentary, pp. 150–58.

97. This notion could result from either direct observation or *Problemata Physica* XV, 6, where we read: "αἱ μὲν εἰς τὰς γωνίας ἀποσχιζόμεναι τῶν ὄψεων ... οὐχ ὁρῶσι," and "Visiones quae ad angulos quidem discinduntur ... non cernunt."

98. Lit.: "capable."

فلو كان الضوء الذي يظهر من ضوء القمر هلاليًا لكان إذا بَعُدَ عن الثقب بَطُلَ تقعيرهُ، لأن أول ما يبطل من الضوء إذا بَعُدَ هو زواياه وحواشيه. فلو كان الذي يُظْهِرهُ من ضوء القمر من القرب هلاليًا، لكان إذا بَعُدَ عن الثقب كان أول ما يبطل منه زواياهُ، ثم حواشيه، فكان يصير مستديرًا. فإذا كان يظهر مستديرًا، وهو قريب من الثقب، فهو إذا بَعُدَ كان أشد استدارة. °||

والبعد الذي نسبته الذي يوجب أن يظهر ضوء القمر الذي يخرج من ثقب مقتدر عندهُ هلاليًا، إذا انتهى ضوء القمر اليه يكون قد تلاشى وبطل، لأن كل ضوء يخرج من ثقب فإنه، إذا بَعُدَ عن الثقب بُعدًا كثيرًا، بطل. °||

وكذلك ضوء الشمس، إذا خرج من ثقب واسع مستدير فإن له بُعد يوجب نسبته أن يظهر الضوء عنده هلاليًا. إلا أن ذلك البعد يكون متفاوتًا، وربما ⌜omL لم B100v توجد على وجه الأرض موضع بُعدهُ من الثقب ⌝omL البعد الذي نسبته النسبة المفروضة. وإن L33r وجد موضع بهذه الصّفة فإن المعتبر، الذي هو عند الثقب، لا يدرك بصره ما يكون في ذلك الموضع لتفاوت بُعدهُ. ومع ذلك فإن ضوء الشمس الذي يخرج من الثقب الواسع،

---

692 بَعُدَ] بعُد B بعد OL || 693 بَعُدَ] بعُد B بعد OL || 694 الذي] om L || 694 يُظْهِرهُ] يظهرهُ B يظهر OL || 694 إذا] om L || 694 بَعُدَ] بعُد B بعد OL || 695 ما يبطل] om L || 695 زواياهُ] زاويتاه BO || 695 فكان] وكان B فكان OL || 696 بَعُدَ] بعُد B بعد OL || 699 مقتدر] مقتدرة B مقتدر L || 699 تلاشى] تلاشا BO || 700 بَعُدَ] بعُد B بعد OL || 702 إلا أن] إذا كان O || 702 يكون] post بُعدًا add B || 703 لم توجد على وجد الأرض موضع بُعدهُ من الثقب] om B || 703 البعد الذي] بما ... توجد L || 703 البعد] الأ من B || 703 الأرض] om B || 703 توجد على وجد] om L

aperture reaches that place, it would vanish and fade away. And this notion <also> concerns lights that pass through narrow apertures. |[O]

<Conclusion>

Thus we have discovered the cause for which the sunlight appears crescent-shaped when, at the time of the eclipse, it passes through one of these apertures and attain a plane parallel to the aperture. <Instead,> the moonlight emerging from these apertures does not appear crescent-shaped [O86v] <neither> when the Moon is crescent-shaped <nor> at the time of the eclipse. Its light coming from these apertures will always appear circular. Moreover, we have shown when the light of the eclipsed Sun appears crescent-shaped and when it appears circular at the time of the eclipse. These are the notions that we intended to explain in this epistle. |[O]||[B]

The end of the epistle.[99]

99. OBL differ from then on: O "Praised be Allah, Lord of the Worlds, peace and blessings upon Muḥammad and all his family." B "All praise be to God, I copied it from a manuscript in his own handwriting, God have mercy upon him. The conformity [of the copy?] with the original was done." L "The end of the epistle by one of the Moderns—lit.: newcomers—and His support."

إذا انتهى إلى ذلك الموضع بَطُلَ واضمحل. وهذا المعنى يظهر من الأضواء التي تخرج من الثقوب الضيقة. |O

فقد تبينت العلة التي من أجلها يظهر ضوء الشمس في وقت كسوفها، إذا خرج من ثقب من الثقوب وصار إلى سطح مواز للثقب هلاليًا. ولا يظهر ضوء القمر، إذا خرج من الثقوب هلاليًا O86v في وقت كون القمر هلالًا وفي وقت كسوفه. فيظهر ضوءهُ الذي يخرج من الثقوب أبدًا مستديرًا. وتبيّن مع ذلك متى يظهر ضوء الشمس إذا انكسفت هلاليًا، ومتى يظهر مستديرًا في حال كسوفها. وهذه هي المعاني التي قصدنا لنبينها في هذِه المقالة. B||O|

تمت المقالة.

---

708 تخرج] يخرج O نخرج L the plural الأضواء requires agreement in fem sing || 710 مواز] موازي BO || 712 ضوءهُ] ضوء L || 712 الثقوب] الثقب L || 713 انكسفت] انكشف L || 714 لنبينها] تبينها B لتبينها O لثبينها L || 716 المقالة] والحمد لله رب العالمين والصّلاة على محمّد واله أجمعين وسلم add O والحمد لله حق حمده نقلتها من نسخة بخطه رحمه الله بلغ علي فصله] مقابلة add marg lectio incerta] add B تمت المقالة بين افّد وتوفيقه add L

Chapter 3
Ibn al-Haytham's Method

This Chapter covers the subjects of *On the Shape of the Eclipse* from a methodological viewpoint. The comment is divided into two main sections: Ibn al-Haytham's Predecessors and Instrument.

1. Ibn al-Haytham's Predecessors

Ibn al-Haytham's own contribution to the problem of the image formation can be highlighted by first recalling what his predecessors already knew about this topic and which approach they developed.

1. Pseudo-Aristotle

The history of the camera obscura (in the Mediterranean) can been traced back to the pseudo-Aristotelian *Problemata Physica*, which address several issues similar to that of *On the Shape of the Eclipse*. We read in Book XV, 6 (911b3):

> "Why does the Sun penetrating through quadrilaterals form not rectilinear shapes but circles, as for instance when it passes through wicker-work? Is it because the projection of the vision is in the form of a cone, and the base of a cone is a circle, so that the rays of the Sun always appear circular on whatever object they fall? For the figure also formed by the Sun must be contained by straight lines, if the rays are straight; for when they fall in a straight line on to a straight line, they form a figure contained by straight lines. And this is what happens with the rays; for they fall on the straight line of the wicker-work, at the point where they shine through, and are themselves straight, so that their projection is a straight line. But because the parts of the vision which are cut off towards the extremities of the straight lines are weak, the parts of the figure about the angles are not seen; but what there is of straight line in the cone describes a straight line, while the rest does not, but the sight falls on part of the figure without perceiving it" (Aristotle 1995: 1417).

D. Raynaud, *A Critical Edition of Ibn al-Haytham's On the Shape of the Eclipse*,
Sources and Studies in the History of Mathematics and Physical Sciences,
DOI 10.1007/978-3-319-47991-0_3

In Book XV, 11 (912b28), the author asks further:

> "Why is it that during eclipses of the Sun, if one views them through a sieve or a leaf—for example, that of a plane-tree or any other broad-leaved tree—or through the two hands with the fingers interlaced, the rays are crescent-shaped in the direction of the Earth? Is it because, just as, when the light shines through an aperture with regular angles, the result is a round figure, namely a cone (the reason being that two cones are formed, one between the sun and the aperture and the other between the aperture and the ground, and their apexes meet) [...] Such crescents are not formed by the Moon, whether in eclipse, or waxing or waning, because the rays from its extremities are not clearcut, but it sheds its light from the middle, and the middle portion of the crescent is but small" (Aristotle 1995: 1419).

Even though Ibn al-Haytham did not make any explicit reference to the pseudo-Aristotelian *Problemata Physica*, Sabra and other scholars surmised that he was aware of it. The text, which was translated into Arabic by Ḥunayn ibn Isḥāq, is known under the title *Problemata Physica Arabica*. It was given an edition by Lou S. Filius from two extant manuscripts: Manisa, Kitapsaray, İl Halk Kütüphanesi MS 1790, fols. 131v–402v, copied in the month of Rajab A.H. 1029/1630; Tehran, University Library, MS 2234, fols. 2r–46r, eleventh A.H./seventeenth century.

The Arabic version presents the following distinctive features.

1. The Greek text copied by Ḥunayn ibn Isḥāq underwent a revision with the effect of putting the handbook in accordance with the doctrines of Galen and Pseudo-Alexander of Aphrodisias. This reworking should have taken place after A.D. 200 (see Filius in Aristotle 1999: XXIV–XXV).

2. The division into chapters slightly differs from the Greek, so that the content of *Problemata Physica Graeca*, Book XV, which anticipates the study of the camera obscura, is found in Book XVII of Ḥunayn's version.

3. Ḥunayn's Arabic version stops after XVII, 2 (i.e., a few lines before the end of *Problemata Physica Graeca* XV, 7), just before switching to the colophon: "Aristotle's book on Physical Questions has been completed with the help of God on the blessed Friday in the month Rajab of the year 1029" (Aristotle 1999: 661). The missing chapters cannot be considered as a subsequent loss, for the text opens with a table of

contents which reads: "Physical Questions of Aristotle by way of introduction to the subject. A translation by Abū Zayd Ḥunayn ibn Isḥāq. Ḥunayn says: This book consist of 17 maqālāt, comprising 372 questions" (Aristotle 1999: LVII; more on this in Filius 2006). The words "Ḥunayn says" tells us that the Arabic version did not go any further; and neither did the *Problemata Physica* translated by Moses ibn Tibbon in the month of Nisan 5024/April 1264 in Montpellier. The Hebrew version stops after the first four chapters of the *Problemata Physica Graeca.*

These facts indeed leave open the possibility that Ibn al-Haytham read the *Problemata Physica Arabica* XV, 6 (which is numbered XVII, 2 in Ḥunayn's version). This Arabic version is quite faithful to the Greek original:

ما بال الشمش اذا جازت على اشكال مربعة الاضلاع لا تفعل اشكالا متساوية الخطوط وتفعل اشكال مستديرة وكذلك ايضا النظر؟ لان النور النظرى فى شكله يشبه قونوس اى السروى وكذلك يكون خروجه وانبساطه وقاعدة قونوس السروى مستديرة. وقد كان يجب ان كان الشعاع مستويا وسقوطه على خطوط مستقيمة ان ترى مستوية على انه من اجل ان الخطوط المتساوية فى مواضع الزوايا ناقصة ضعيفة لا تظهر الزوايا بل ما كان من القونوس داخلا فانه قابل للنور (*PPA* XVII, 2; Aristotle 1999: 658)

Why is it that when the sun passes through quadrilateral forms, it does not produce figures with straight lines, but it does produce round figures, and likewise the sight? Because the light of perception in its form resembles a κῶνος, i.e., cone. The light of perception also emanates and spreads in this way. The base of the κῶνος, the cone, is round. Perhaps it would be necessary for the rays to be straight and for them to fall down in straight lines, if one saw them straight. But because the straight lines in the place of the angles are deficient and weak, the angles are not visibles, but (only) the inside of the κῶνος, for the latter is receptive to light.

The explanation is overly simplistic: the light of the Sun appears round because the light cone is round and the angles are weak. Formulas like "Such crescents are not formed by the moon, whether in eclipse or waxing or waning" inevitably makes us think of Ibn al-Haytham's text. On the the other hand, there also exist significant differences between the *Problemata Physica Arabica* and Ibn al-Haytham's work with regard to scientific vocabulary:

Problemata Arabica	*Ibn al-Haytham*	*Meaning*
الشكل مربع	الثقب مربع	square aperture
النور	الضوء	light
قونوس، السروى	المخروط	cone
انبساطه	الممتد	extension
التباعد	البعد	remoteness

The notion that Ibn al-Haytham built his text upon the *Problemata Physica* thus remains hypothetical: there are obvious similarities between the two texts; however terminology differs widely and the solar eclipse mentioned in *Problemata Physica Graeca* XV, 11, does not even appear in Arabic. Consequently, it remains unclear whether Ibn al-Haytham ever reproduced Pseudo-Aristotle's errors.[10]

Al-Kindī—9th Century

Al-Kindī is the author of the *Liber de causis diversitatum aspectus*—commonly shortened to *De aspectus*—whose Arabic text has been lost and only survives in Latin. The text was first edited by Björnbo and Vogl (1912) and more recently by Hugonnard-Roche (Jolivet and Rashed 1997). The camera obscura comes into play in the course of Proposition 6 (Fig. 3.1):

10. After a reference to the shortcomings of XV, 11 Lindberg states: "This erroneous conclusion was to be reproduced by Ibn al-Haytham in his treatise, *On the Shape of the Eclipse*" (1968: 159). Ibn al-Haytham's dependency on the *Problemata* has also been claimed by Sabra: "The problem had already been posed in somewhat similar terms in the pseudo-Aristotelian *Problemata*, a work which, in some form, was available to I.H. and which he abridged or paraphrased some time before the end of A.H. 417" (Sabra 1989, II: xlix–l). If there is no doubt that Ibn al-Haytham *could* access the *Problemata Physica Arabica*, the fact that he abridged them, a fact known to us from Ibn al-Haytham's list of works included in Ibn Abī Uṣaybiʿa's *Ṭabaqāt*, is controversial. Rashed (1993, 2007) supported that this list mingled the works of Muḥammad the philosopher with those of al-Ḥasan the mathematician. If we were to follow his opinion, the dependency of Ibn al-Haytham's on the *Problemata* would be indefinite. Without engaging in this complicated debate, suffice it to say that the sudden stop of the *Problemata Arabica* after Section XV, 7 casts a doubt on the influence of this text, because one major part of the discussion of pinhole images, appearing in XV, 11, is lacking in Arabic.

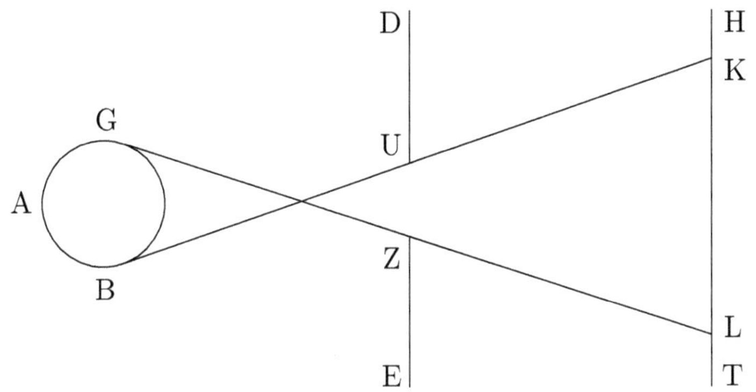

Fig. 3.1. Al-Kindī, *De aspectus*, Prop. 6

Hoc quoque manifestius et clarius videbimus, si tabulam assumpserimus in medio cum serra directe et equaliter perforatam, et occurrerimus deinde cum medio foraminis serrati medio candele, donec sit linea, que protrahitur a candela, secans diametrum candele et foramen serratum orthogonaliter, et post occurramus tabule, in qua est foramen, cum alia tabula, cuius superficies, que ei occurrit, equidistet superficiei eius que ipsi occurrit...

This [the rectilinear propagation of rays] will be still more evident and clear if we take a board and make a hole perpendicularly and regularly in it with a saw; if we then place the centre of the aperture made with a saw and the centre of a candle opposite each other, so that the line drawn from the candle cut orthogonally the candle's diameter and the aperture made with a saw; and finally if we place behind the opened board another board, the surface of which being opposite and parallel to the first board...

Verbi gratia sit candela circulus *ABG*, et sit tabula perforata tabula *DE*, sitque foramen spatium *UZ*. Tabula quoque, super quam cadit lumen, sit *HT* et pars eius, super quam cadit lumen, sit spatium *KL*. Cum ergo protrahetur linea a nota *K* ad notam *U*, et producetur recte, perveniet ad notam candele que est *B*, que est e contrario partis *U*...

For instance, let be circle *ABG* the candle, board *DE* the perforated board, and space *UZ* the aperture. And again, let be *HT* the board on which the light is falling, and let be space *KL* the part of the board the light is falling on. If therefore we draw a line from point *K* to point *U*, and extend it by a straight line, it will fall on point *B* of the candle, which is on the side opposite to *U*...

Quod quidem non esset nisi radiorum fines secundum rectas procederent lineas (Rashed 1997: 449–51).

Definitely, that would not happen if the limits of the rays did not follow straight lines.

Al-Kindī's text is interesting in more than one respect. Firstly, it is clear that he used the camera obscura only for demonstrating thc rectilinear propagation of light. There is no discussion of any image cast onto a screen whatsoever. Secondly, the illuminated part of the screen is bounded by points *KL*, which are aligned with points *BG* of the light source and points *UZ* of the aperture. The text mentions that *B* is "on the side opposite" to *U*. This is the premise of the subsequent observations of the image inversion in the optical literature. Thirdly, the text makes it clear that the aperture was drilled into the board with a saw. The Latin has "tabula cum serra perforata," "foramen serratum". As a result, the aperture was large—at least wider than the saw—a condition that prevents the image to be stigmatic, as we will see in detail in Chapter 4.6, pp. 104–11.

Pseudo-Euclid—?

The *Book of Mirrors of Euclid* is an optical compilation falsely attributed to Euclid. The Greek original is lost. The text is extant in Arabic, Latin and Hebrew.

The Arabic version (*Kitāb al-mir'āh li-Uqlīdis*) is known from only a little fragment, which appears in Florence, Medicea Laurenziana, MS. Or. 152, among a compilation of astronomical tables and tracts (Assemani 1742: 389–90). The Florentine codex consists of two parts bound together in one volume. The first collection (olim 280), written in the same *āndalusī* hand, was completed at Toledo at the court of Alfonso X el Sabio before 20 March 1303 of the Hispanic era, i.e., 1265 A.D. The second part of the book (olim 279) was completed 21 Sha'ban 664 A.H./4 June 1266. The fragment *On Mirrors* appears on fol. 104, just at the end of the first part (Sabra 1977; Rashed 1997: 337; Kheirandish 1999: 243). It is apparent that Prop. 4, beginning with: "Procedit a pupilla virtus luminosa imprimens in eo..." is repeated in the introduction of al-Kindī's *Rectification of Error and Difficulties due to Euclid's Book on Optics* (*Kitāb fī taqwim al-khaṭa' wa al-mushkilāt allatī li-Uqlīdis fī kitābihi al-mawsūm bi-al-manāẓir*). This can be explained in two ways: either the former copied from the latter, or the reverse. As the *Rectification* pursues a single goal—the critical scrutiny of Euclid's *Optics*—the principle of parsimony suggests that the *Kitāb al-*

mir'āh li-Uqlīdis came after the *Rectification* (on this matter, see also Björnbo and Vogl 1912: 157).

The Latin version (*Tractatus Euclidis de speculis*) is extant in more than eighteen manuscripts, some of the oldest of which are Oxford, Bodleian, MS Digby 40, fols. 13r–15v, 13th c.; Cambridge, University Library, Ii.1.13, fols. 40r–41r, dated 1279; and Paris, BnF, MS. Lat 9335, fols. 82r–83v, 13th c. Thirteen manuscripts we used to prepare the first scholarly edition (Björnbo and Vogl 1912).

The Hebrew version (ספר המראים, *Sefer ha Mar'im*, that is, *Book of Mirrors*) is, as far as I know, extant in four manuscripts, none of which has a date of composition (Steinschneider 1893: 520–2; 1956: 512–3). Although the translator of this work is unknown, there are grounds for believing that it was redacted in 13th-14th-century Provence (see Lévy 1997: 433).

A proto camera obscura is described at Props. 9 and 10 of this tract. Let us have a closer look at these propositions (Björnbo and Vogl 1912: 102–3, Fig. 3.2):

Et hoc quoque ostendam, quod, cum sol intrat per fenestram, illud luminis eius, quod ingreditur et super terram cadit, magis est amplum quantitate fenestre.	This also will show that, when the sun enters through a window, the light of it, which goes in and falls upon the Earth, is in proportion to the size of the window.
Signabo igitur circulum solis, qui sit circulus *A*, et imaginabor egressus radiorum ab eo, quemadmodum descripsimus in figura, que est ante hanc, et formabo fenestram, que sit introitus, qui est inter *A* et *B*. Et sit superficies terre linea *TZ*. Et producam radium *G* a sole, secundum quod prediximus, et transeat per punctum *A* ab extremitate fenestre, et cadat super punctum *T* superficiei terre.	Therefore, I will draw the Sun as circle *A* and rays emerging from it will be imagined as described in the previous figure, and I will trace the window, by which it enters, between points *A* and *B*. And let the surface of the Earth be line *TZ*. Then I will extend ray *G* from the Sun, as said before, that will pass through point *A* on the edge of the window and through point *T* on the surface of the Earth.
Iam igitur manifestum est, quod linea *TZ* maior est introitu *AB*. Linea autem *TZ* est, quod contingit radius ex superficie terre, et introitus eius per spatium *AB* est quantitas amplitudinis fenestre.	Then it is clear that line *TZ* is greater than aperture *AB*. Line *TZ* is the limit of the ray on the face of the Earth, and its entrance through interval *AB* is the size of the window.

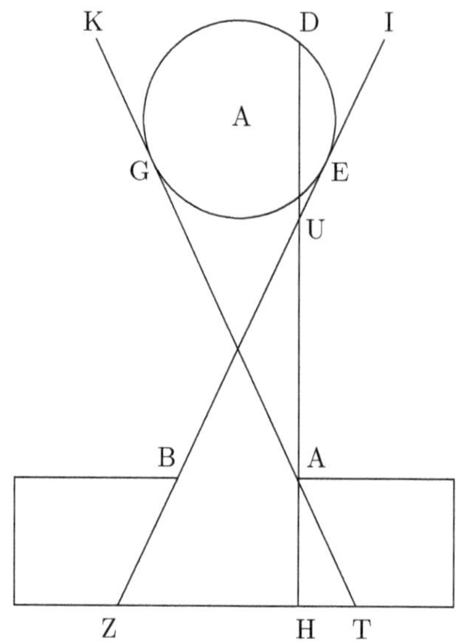

Fig. 3.2. Pseudo-Euclid, Props. 9–10

Cum autem sol eclipsatur, tunc illud luminis eius quod per fenestram ingreditur diminutum est, scilicet non rotundum, neque casus eius super superficiem terre est sicut linea *TZ* sed est secundum quantitatem eius eclipsis; et est diminutio luminis proportionalis diminutioni eclipsis.

Verbi gratia quasi sol eclipsetur, et sit eclipsis eius arcus *DGU*, et sit pars corporis eius, que prohicit radium tunc, arcus *DEU*, et ingrediatur radius arcus *DEU* per fenestram ad superficiem terre, que est linea *TZ*, linea *TH*. Et deest ad complementum radii supra superficiem terre linea *HZ*; ipsa namque est, que illuminatur ex radio arcus *DGU*. Ergo proportio linee *HZ* ad arcum *DGU* est sicut proportio linee *TH* ad arcum *DEU*. Hoc quoque similiter invenitur, secundum quod posuimus et premisimus in omni loco et omni tempore, cum hoc accidens contingit. Et illus est, quod demonstrare voluimus (Björnbo 1912: 102–3).

When the Sun is eclipsed, the light of it, which enters through a window, is diminished, i.e., not a round, neither its appearance on the surface of the Earth is such as line *TZ*, but it is according to the size of its eclipse; and the diminution of light is proportional to the diminution of the eclipse.

Consider e.g., the eclipsed Sun, and let its eclipse be arc *DGU*, and the part of its body expelling the ray be arc *DEU*. Let arc *DEU* be passed through the window to the surface of the Earth, which is line *TZ*; this will be line *TH*. The remainder of the ray [which] is missing on the surface of the earth, will be line *HZ*, for it is the part which is lit up by the ray from arc *DGU*. Therefore the ratio of line *HZ* to arc *DGU* is as the ratio of line *TH* to arc *DEU*. The same will be found in any place and at any times, according to the hypotheses and premises, when this happens accidentally. This is what we wanted to prove.

Basically, the author of the *Kitāb al-mir'āh li-Uqlīdis* follows the same reasoning and arrives at the same conclusion as al-Kindī. He uses the same diagram, except in that the screen is replaced by a box *ABZT* and line *DH* is drawn. There is no longer mention of how the aperture is drilled into the front face, and hence the size of the aperture *AB* can be anything. The main difference between the two tracts is that the reasoning is applied to a partial eclipse of the Sun (Prop. 10).

Al-Khujandī—10th Century

The Central Asian astronomer al-Khujandī is famous for the giant mural sextant he built for the Būyid emir Fakhr al-Dawla. This instrument, which was designed to measure the obliquity of the ecliptic and local latitude, was installed on Mount Ṭabrūk, near Rayy. No vestiges of the building remain. The instrument is only known from al-Khujandī's text (Cheikho 1908) and two further comments by Bīrūnī (1967: 70–77) and al-Marrākushī (Sédillot 1844: 202–206). Despite its names—mural sextant, aperture gnomon—al-Khujandī's instrument was a kind of dark chamber:

> "The aperture gnomon arose at some point in Islamic astronomy, when people realised the great advantage of replacing the shadow of the tip or edge of the gnomon by the cone of light emerging through the tiny aperture, in the manner of a camera obscura [...] The earliest medieval attribution of the use of this instrument refers to al-Khujandī" (Mercier 1984: 162).

The Rayy gnomon had a radius of 40 *dhirā'* (21.6 m) and an aperture of half a *shibr* (12.4 cm) after al-Bīrūnī, or 1 *qabḍa* (9 cm) after al-Marrākushī (Sayılı 1960: 118–21; Oudet 1994: 36; by setting the cubit to 532.2 mm, p. 222). The sunlight was cast onto a moving board, which was driven along the two parallel meridian arcs.

At later dates, similar instruments were built in Marāgha and Samarkand. Ulugh Beg's sextant had a radius of 75 *dhirā'* (40.2 m) and an aperture of 1 *qabḍa* (9 cm) (Oudet 1994: 45–8). These dimensions together with the orbital elements allow us to calculate the dimension of the patch of light on the board. Let *AB* be the diameter of the Sun, *GD* the diameter of the aperture and *EH* the projection plane. As the Sun is virtually at infinity, we draw *GT* parallel to *DZ* (Fig. 3.3).

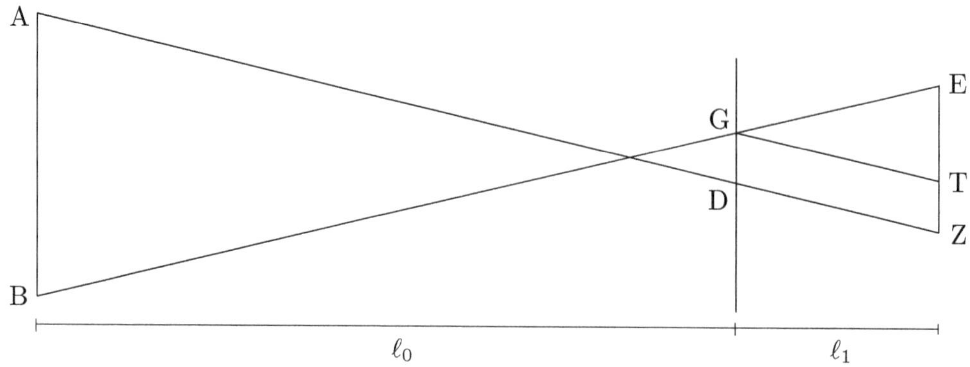

Fig. 3.3. The Aperture Gnomon by al-Khujandī

The patch of light EZ consists of two parts: the penumbra TZ, which I represent on one side only, though it stands on both sides of the image, and the central area of full light, which is $ET - TZ$. We know that $ET = AB\,\ell_1/\ell_0$ and $TZ = GD$.

Irrespective of the case, $\ell_0 = 1$ AU.

Therefore, all elements are known and the image provided by the mural sextant or gnomon may be reconstructed in each situation (see Plate 3.3, p. 250):

R1: According to al-Bīrūnī, in Rayy, $\ell_1 = 21.6$ m and $GD = 124$ mm; therefore the gnomon generated a patch of light $124 + 69 + 124 = 317$ mm wide.

R2: According to al-Marrākushī, in Rayy, $\ell_1 = 21.6$ m and $GD = 90$ mm; thus the gnomon generated a patch of light $90 + 111 + 90 = 291$ mm wide.

S1: In Samarkand, $\ell_1 = 40.2$ m; therefore the instrument generated a patch of light $90 + 284 + 90 = 464$ mm wide.

The visual inspection of the diagrams shows that, even though al-Khujandī's gnomon was very similar to a camera obscura in its design, it provided no sharp image because the penumbra border TZ caused a blurring of the image.

However, the instrument designed by al-Khujandī displayed characteristics worthy of interest for the history of optics:

1. Firstly, the measurement of meridian transits assumed a real image of the Sun to be cast onto the projection plane, i.e., the board moved on parallel meridian arcs.

2. Secondly, though still wide (one *qabḍa*), the opening of al-Khujandī's gnomon was reduced in size compared to al-Kindī's instrument, thus anticipating future advances on the camera obscura problem.

'Uṭārid al-Ḥāsib—10th Century

I will touch only briefly on the 10th-century astronomer 'Uṭārid ibn Muḥammad al-Ḥāsib, who commented on the work on burning mirrors by Anthemius of Tralles, the celebrated architect of Hagia Sophia. His text, titled *The Bright Lights for the Construction of Burning Mirrors* (*al-Anwār al-mushriqa fī 'amal al-marāyā al-muḥriqa*), is kept in only two manuscripts (Istanbul, Süleymaniye kütüphanesi MS Laleli 2759, fols. 1v–20r, written before 899; Ayasofya, MS 2676, fol. 1v–18v, no date). This work is worthy of note here, for it has been claimed that it included "the earliest occurrence of *al-bayt al-muẓlim* (camera obscura)" (Sabra 1972: 204, n. 19). As 'Uṭārid's revision is now available in scholarly edition (Rashed 1997: text 296–315), I have looked at and found that the word بيت المظلم (camera obscura) does not appear once in the text.

What, instead, is true, is that this tract includes a vague notion of darkroom for it is said at the beginning of the text that a "ray passes through an aperture." The text does not actually state that the ray arrives in an enclosed area like a cubbyhole. Another reason why there is no need to proceed any further with this source is that Uṭārid's and Ibn al-Haytham's terminologies are completely different, as can be seen from the following words:

'Uṭārid	*Ibn al-Haytham*	*Meaning*
نمدّ	نصل	we draw
ينفذ	يمر	it passes
الشعاع	الخط	ray
الكوة، الخرق	الثقب	aperture
قبالة	مقابل	opposite

2. The Archimedean Analysis

Although Archimedes is not referred to by name in *On the Shape of the Eclipse*, Ibn al-Haytham's work inherited two important Archimedean legacies well disseminated in Arab medieval mathematics: the first one involves mechanical means or instruments in the demonstration, the second is reasoning on infinitesimal quantities (see Youschkevitch 1967; Clagett 1964–80).

On the Shape of the Eclipse sets out to determine the shape of the image by a mechanical means, involving a straight line joining a point of the Sun to the projection plane (see Section 3.4: *The Archimedean Analysis*, lines 348–430). While the line is driven along the perimeter of the aperture, the endpoint of the line traces the perimeter of the image onto the projection plane. The shape of the image is thus the result of a hybrid method, halfway between geometry and mechanics. This is not pure geometry, which does not allow any geometrical proof to be derived from a construction by means of instruments and motions. Several attempts to include such deviant methods in the realm of pure mathematics were made in Greek Antiquity and Medieval Islam. Archimedes advocated this alternative approach in the introduction of the *Method:*

> "Seeing moreover in you, as I say, an earnest student, a man of considerable eminence in philosophy, and an admirer of mathematical inquiry, I thought fit to write out for you and explain in detail in the same book the peculiarity of a certain method, by which it will be possible for you to get a start to enable you to investigate some of the problems in mathematics by means of mechanics. This procedure is, I am persuaded, no less useful even for the proof of the theorems themselves; for certain things first became clear to me by a mechanical method, although they had to be demonstrated by geometry afterwards because their investigation by this method fell short of demonstration. But it is of course easier, when we have previously acquired, by the method, some knowledge of the questions, to supply the proof than it is to find it without any previous knowledge" (Heath 1912: 13).

Archimedes used this method for generating the spiral in *On Spirals*, for squaring the parabola in the *Method of Mechanical Theorems*. He also gave a *neusis* construction for the regular heptagon in the *Book on the Construction of the Circle Divided into Seven Equal Parts.* Eutocius of Ascalon (fl. 540 CE) also informs us in his *Commentary on the Sphere and Cylinder* II, 1, that other Greek mathematicians obtained two proportional means by using mechanical means. Most of them used a "shifting ruler" (κανών) (such as Hero, Philo, Apollonius, Diocles and Pappus of Alexandria) or a "peg" (τυλόν) driven in a "groove" (σωλήνας) (such as Hero, Pappus and Nicomedes), less frequently an instrument equipped with "moving tablets" (πινακίσκος) (Eratosthenes) (Eutocius 1972: 45–56 and 62–75).

Despite the fact that Archimedes was criticized by al-Qūhī, al-Sijzī and others for leaving the lemma on the side of the regular heptagon undemonstrated, Arab scholars used extensively mechanical means. This was indeed the case of Aḥmad Ibn Mūsā, Thābit Ibn Qurra, Abū al-Wafā' and, at a later date, Badr al-Dīn Muḥammad (see Hogendijk 1984: 280–281; Knorr 1989: 213–219). By resorting to mechanical geometry—*continuously* moving the apex of a light cone along the perimeter of the aperture to create an *infinity* of patches of light overlapping each other on the projection plane—Ibn al-Haytham took place along these lines. Introducing infinitesimal quantities is another key feature, which we will be analysing below.

3. The Point Analysis of Light

The scientific achievement that distinguishes *On the Shape of the Eclipse* is mainly due to the combination of experimental reasoning and geometry. One of its special features is the resort to the point analysis of light, a term that physicist and science historian Vasco Ronchi coined to characterize Ibn al-Haytham's theory of vision:

> "With this approach Alhazen overcame the two main obstacles which, as we remarked at the beginning of this chapter, had stopped all further development of previous theories concerning vision... He divided the visible object into point-like elements and in this way he suppressed the character of an indivisible and total operation which until then was attributed to the act of seeing the object [...] Alhazen had put forward the ingenious idea of considering an object as being made up of small elements each of which sent out its point-like *species* along a ray until it entered the pupil and reached the crystalline lens" (Ronchi 1970: 50, 83).

Al-Kindī marked a first milestone in this history with his *Liber Jacob Alkindi de causis diversitatum aspectus*, often abbreviated to *De aspectus* (Rashed 1997: 485–7; Fig. 3.4). In the course of Prop. 14, we read:

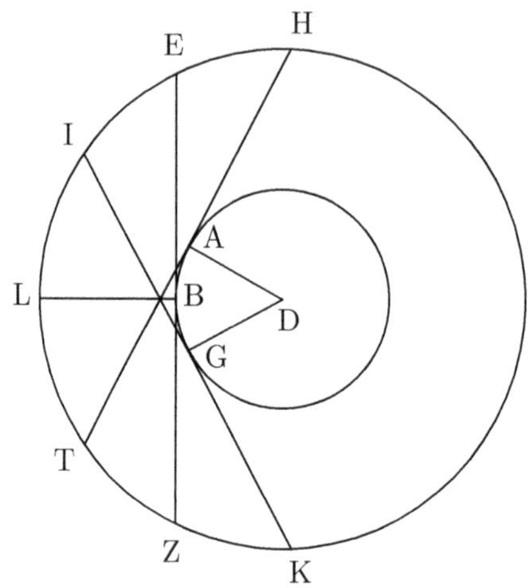

Fig. 3.4. Al-Kindī, *De aspectus*, Prop. 14

Dico ergo quod locus, qui obviat pluribus partibus luminosis, plus illuminatur eo qui paucioribus occurrit...

Cuius exempli causa sit instrumentum visus, scilicet oculus, circulus *ABG*, cuius centrum est *D*. Et sit pars, cui inest potentia comprehendendi visibile, videlicet dicta exterior gibbositas oculi, arcus *ABG*... Et sit corpus, quod conspicitur, arcus *HEITZK*, et sit nota, que subest centro visus, nota *L*, a qua cum producitur linea ad *B*, sit perpendicularis supra lineam *EZ*...

Pars ergo *L* illuminatur a tribus partibus *A*, *B* et *G*.

Arcus autem *EI* est communis duobus arcubus simul *HT* et *EZ*. Ipse igitur illuminatur simul a duabus partibus *A* et *B* tantum...

Arcus vero *HE* est pars arcus *HT* solum. Non ergo illuminatur nisi a parte *A* tantum (Rashed 1997: 485–7).

I say that the place facing more luminous parts is more illuminated than that which is before fewer <parts>...

For example, let the organ of vision, that is to say the eye, be circle *ABG* of center *D*. And let the part in which is the power to capture the visible, i.e., the part which is called the convexity of the eye, be arc *ABG*... And let the body which is viewed be arc *HEITZK*, and let the point at the center of sight be point *L*, from which the line toward *B* is drawn perpendicular to line *EZ*...

Therefore part *L* is lit up from the three parts *A*, *B* and *G*.

Arc *EI*, on the other hand, is common to both arcs *HT* and *EZ*. It is thus illuminated from both parts *A* and *B* only...

As to arc *HE*, it is a part of arc *HT* only. Therefore it is not lit except from part *A* only.

The segmentation of the light source into small elements gives us information about the illumination of any body opposite to the light source: the greater the number of rays falling on a point of this body, the more illuminated the point. Clearly, this is an anticipation of the point analysis that can be found in *On the Shape of the Eclipse*. However, al-Kindī did not go as far as Ibn al-Haytham, for two reasons:

Firstly, in Prop. 14, al-Kindī is not wondering about the variation of natural light. Instead, he is considering the variation of the visual faculty emanated from the eye and received on the visible body HEITZK, because he was a proponent of the theory of extramission. Note, however, that his proof, which is attached to the extramission, can be easily converted to physical optics provided that the visual faculty is equated to natural light (point D shall be construed as a light source, arc HEITZK as a screen, and the like). Once these premises are accepted, the result is the same: LI is lit thrice, IE twice, EH once... Therefore LI will be more illuminated than IE and IE more than EH.

Secondly, al-Kindī's *De aspectus* lacks any reference to the continuous dimming of light or, what amounts to the same, any mention that arc HEIL can be segmented in as many segments as one wishes. Therefore there is no mention whatsoever of infinitesimal quantities in the *De aspectus*, as is the case in Ibn al-Haytham's work.

In spite of these shortcomings, al-Kindī contributed significantly to the analysis of the formation of images in applying a point analysis of light.

Ibn al-Haytham approached the formation of images in the same manner. He postulated that an image is composed of elementary patches of light, each generated by a ray projected through the aperture. This analytical scheme was first applied in *On the Qualities of Shadows* (*Maqāla fī kayfiyyat al-aẓlāl*), a work on the penumbra. This work was translated into German (Wiedemann 1907) and benefited from a comment (Naẓīf 1942: 264–274). Ibn al-Haytham first defines two concepts for the study of the penumbra: *al-ẓill al-maḥḍ* or *ẓulma* ("pure, unmixed shadow" or "darkness," i.e., the total absence of light) and *al-ẓill* ("shadow," i.e., the partial absence of light). He then applies a point analysis of light to the gradual shadow ḤH, which is cast by the opaque line ǦD when it is illuminated by the light source AB (Fig. 3.5).

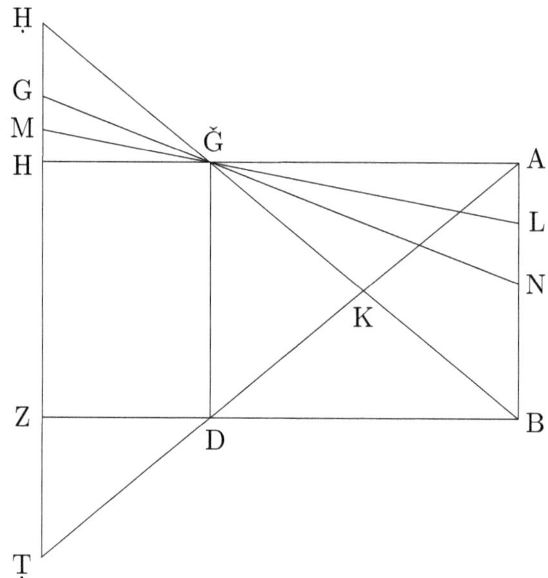

Fig. 3.5. Ibn al-Haytham, *On the Qualities of Shadows*

"From every part of the luminous body *AB* light goes out in straight lines. Let this part be *AL*. We draw *LǦ* that cuts *ḤH* at point *M*. All the lines extended from a point of *AL* to *Ǧ* cuts line *HM* ... thus *HM* receives light from segment *AL*. By contrast, any straight line produced from a point of *LB* to *HM* is stopped by *ǦD*. So *HM* does not receive some of the lights from line *AB*. Consequently, there is a mix of light and shade in line *HM*. Moreover, line *AL* is much smaller than *AB*, so there is far more shade than light in *HM* ... Then we make *LN* = *AL*, we draw *NǦ*, which meets *ṬḤ* in *G*. The light of *AL* falls on *MG* too, so that line *MG* is lit twice by the light of the two segments *AL* and *LN* ... Therefore light is stronger on *MG* than on *HM*, and the shadow is weaker on *MG* than on *HM*. It is clear that, in line *HḤ*, the shadow is continuous (*ẓill muttaṣil*) and nonetheless variable (*wa-ma'a dhalika mukhtalif*): close to point *H* it is stronger; close to point *Ḥ* it is weaker ... Thus in *HḤ*, the shadow is variable and the variation is gradual (*wa-ikhtilāf 'alā tadrīj*), with no separation from one part to another. Similar considerations apply to *ZṬ* ... By contrast, line *ZH* is shrouded in darkness, because its points receive no light from *AB*. If a line is drawn from it *[AB]* to a point of *HZ*, it will be stopped by *ǦD*" (Fātiḥ 3439, fol. 124v–125r).

Ibn al-Haytham's reasoning is clear. Let us cut the light source AB into as many equal parts as desired, say, four parts, to follow the diagram. HM receives light from AL/AB, that is 1/4 of the source; MG receives light from (AL+LN)/AB, that is 2/4 of the source, etc., GḤ receives light from (AL+LN+NB)/AB, that is 4/4 of the light source. Therefore point Ḥ is brighter than G, point G is brighter than M, and point M is brighter than H. Ibn al-Haytham goes one step further in saying that the shadow exhibits a gradual variation (*wa-ikhtilāf 'alā tadrīj*) along *ḤH*, which is "continuous" (*muttaṣil*) and "with no separation from one part to another."[11] In short, *On the Qualities of Shadows* establishes the continuous variation of light in the penumbra from full light, in Ḥ, to full shade, in H.

How are we to understand this outcome from the steps of the demonstration? A first observation is that neither the length of the segment AL nor the ratio AL/AB is defined in the text, which simply states: "line AL is much smaller than AB" and "we make LN = AL." Accordingly, the light source can be cut into as many segments as one wishes. If the length of the segments tend towards zero, they are infinitesimal quantities. This perspective is the only one able to explain why the shadow should vary continuously. The use of infinitesimals is, without being totally explicit, needed to achieve the outcome *On the Qualities of Shadows*. It is worth noting that a similar analysis of light resurfaced at later dates in the works of Leonardo Da Vinci and Francesco Maurolico (for details, see Raynaud 2012*b*).

4. Ibn al-Haytham' Experimental Method

Science historians have already noted that Ibn al-Haytham used the experimental method (Naẓīf 1942–3, I: 43–48; Schramm 1963; Sabra 1971; Omar 1977; Sabra 1989, I: 14–19; Gorini 2003; Kheirandish 2009). In this respect, it is important to note that Ibn al-Haytham did not use the word *tajriba*, which was the standard Arabic translation of the Aristotelian word εμπειρία (Janssens 2004). He used instead a term which denotes a test proceeding with the comparison of different sets of data. This adjusts perfectly with the current definition of experimentation.

11. The word *muttaṣil* already means in itself "uninterrupted."

Claude Bernard is acknowledged as the author of the most commonly used definition of the experimental method. In *Introduction à l'étude de la médecine expérimentale* he opposed the French naturalist Georges Cuvier, for whom the difference between observers and experimenters relied primarily on the criterium of activity. Bernard understood that this criterium was sometimes insubstantial, sometimes contradicted by the facts, and that is why he replaced it with that of modification:

> "We give the name observer to the man who applies methods of investigation, whether simple or complex, to the study of phenomena which he does not vary and which he therefore gathers as nature offers them. We give the name experimenter to the man who applies methods of investigation, whether simple or complex, so as to make natural phenomena vary, or so as to alter them with some purpose or other, and to make them present themselves in circumstances or conditions in which nature does not show them" (Bernard 1999: 15).

This methodological character is resolutely upheld in *On the Shape of the Eclipse.* The text includes six items of *i'tabara* (to experiment, lines 28, 671_1, 671_2, 673_1, 673_2, 680), one item of *mu'tabir* (experimenter, line 705) and one item of *i'tibar* (experimentation, line 47), which—contrary to a thought experiment—calls for control over real conditions. Ibn al-Haytham writes for instance: "If one experiments sunbeams issuing from wide apertures..." (line 28), "The aperture by which the sunlight is tested..." (line 671), and the like.

Ibn al-Haytham varies whatever can be changed: the luminary (the Sun vs. the Moon); the shape of the aperture (circle vs. square); the size of the aperture (narrow vs. wide); the focal distance (near vs. far away). In that respect, most of *On the Shape of the Eclipse* is dedicated to accounting on these experiments.

Beyond the geometrical analysis of the formation of real optical images, Ibn al-Haytham's work provides undeniable evidence of the efficiency of the experimental method to generate knowledge.

First of all, by the use of the experimental method, Ibn al-Haytham is clearly distinguishable from his predecessors such as Aristotle, al-Kindī or al-Khujandī, who all used the dark chamber as a simple observation tool. Ibn al-Haytham thus succeeded in solving long-standing problems.

Second, it should be noted that Western science took a significant amount of time to rediscover Ibn al-Haytham's results. The figureheads of medieval optics, such as Bacon, Pecham and Witelo, generally failed to explain the shape of the light in the camera obscura. Only two authors, Egidius de Baisiu and Levi ben Gerson, get close to Ibn al-Haytham's own results. However, they did not apply the systematic approach with which Ibn al-Haytham succedeed in studying the problem of the formation of images in the camera obscura.

Despite its title, Ibn al-Haytham's *On the Shape of the Eclipse* is much less an astronomical work than an optical treatise, for the solar eclipse dealt with is rather an excuse to develop an in-depth study of the formation of the image, which is analytical in nature.

5. Ibn al-Haytham's Device

The dark chamber (*al-bayt al-muẓlim*) used to observe the partial eclipse of the Sun is not described *expressis verbis* in the text. However, Ibn al-Haytham certainly used a simple, solid darkroom as we are going to see further on.

1. Purposes of the Camera Obscura

As a rule, an experimental device should not be considerated separate from the purpose for which it is designed. In that respect, all darkrooms are different; some properties are critical here, superfluous there. For instance, it is known that the mural sextants of Rayy, and later Samarkand, in which the solar disc was shown on a moving board, were based on the principle of the camera obscura (Sédillot 1844; Oudet 1994). This darkroom was used to determine the latitude and the obliquity of the ecliptic, not to produce an image of the Sun. By comparison, the camera obscura used by al-Kindī was intended to prove the rectilinear propagation of light by examining the boundary between light and shadow—not to produce an image of the Sun. In Ibn al-Haytham's *Kitāb al-manāẓir*, the dark chamber has a broad range of applications, which are, after Sabra's numbering, the following:

I,3 [3–6]: Light propagates in straight lines as shown in dark chambers;

I,6 [85]: Light rays intersecting neither blend, nor mingle, within the pinhole;

I,6 [95]: Light propagates in straight lines, as shown by interposed bodies;

I,8 [3]: To be visible, a body must be separated from the eye;

II,3 [60–61]: Perception of light and color requires time;

II,3 [80]: The distance of a body is assessed by means of interposed objects;

...

Such diversity prompts us to distinguish between all darkrooms from their specific purposes. This should be kept in mind in the course of the discussion.

2. Shape of the Camera Obscura

Let us now describe the dark chamber appearing in *On the Shape of the Eclipse.* When the sunlight (*ḍaw' al-shams*) illuminates a narrow aperture (*thaqb ḍayyiq*) pierced into the front face of the darkroom, it is then cast on the rear face of the chamber. Ibn al-Haytham called this manifestation "the image of the sunlight" (*ṣūra ḍaw' al-shams*), which is not necessarily a stigmatic image in the sense of modern optics. Ibn al-Haytham used this device to fully experiment on the image by varying the luminary, shape, size and distance of the pinhole.

Ibn al-Haytham's darkroom was a cubic box, with vertical parallel faces—probably a simple room surrounded by walls. The front face was pierced by an aperture, which is sometimes specified as a narrow circular aperture. The rear face, which served as projection plane or screen, was plane and parallel to the front face. The most critical aspects of this description could be verified in a few steps.

Walls. Wiedemann (1914: 155–169) translated *saṭḥ* as "Wand," and so did Naẓīf (1942: 186) and Sabra (1989, I: 13–51) in rendering it as "wall". Moreover, as Sabra noted, Ibn al-Haytham sometimes used *jidār* (wall) instead of *saṭḥ* (surface). Thus it comes as no surprise that subsequent scholars used a chamber with vertical walls (see Plates 3.1ABC, pp. 246–8)[12]. The model of Ibn al-Haytham's camera obscura held by

12. For those who are not familiar with the history of the camera obscura, here are some milestones: Alhacen Latinus (12th c.), Alhacen Italicus (14th c.), Guillaume of Saint-Cloud (1292), Egidius de

the Institut für Geschichte der Arabisch-Islamischen Wissenschaften in Frankfurt (Optics, Inventory No. E 2.01; Sezgin and Neubauer 2011: 182–184) is a simple cubic house with parallel vertical walls. This is backed up by several elements.

Verticality. Throughout *On the Shape of the Eclipse*, the solar crescent-shaped image is said to be made of two interlocking equal circles (*dā'iratayni mutasāwyatayni*): "And the two arcs of the crescent are two equal circles, because the light crescent consists of two equal circles."[13] We can be confident that the crescent consisted of circles rather than ellipses because Ibn al-Haytham, as a mathematician, had a complete mastery of conic sections. The *Conics* of Apollonius are even quoted in *On the Shape of the Eclipse* to prove the crescent-shaped sun and the crescent-shaped image to be homothetic with each other: "Any plane, cutting the body of the cone parallel to the base, has... a similar shape to the base of the cone, as shown in Book I of *Conics*."[14] In this context, there can be no confusion between the circle and the ellipse and it is certain that the image was undistorted.

Movability. There are two ways to explain why the solar image was circular. Either the dark chamber was a movable device that could track the sun, or it was a fixed device and the eclipse occurred under particular conditions. The text explains:

> "The plane of the aperture is parallel to the plane of the circle tangent to the first cone along the perimeter of the Sun, or nearly. Between them, namely, the plane of the aperture and the plane of the circle bounding the Sun, there is no perceptible <angular>

Baisiu (ca. 1300), Gersonides (1329), Da Vinci (ca. 1490), Maurolico (1521), Don Papnutio (Cesariano 1521), Reinhold (1542), Gemma Frisius (1545), Cardano (1550), Della Porta (1558), Santbech (1561), Barbaro (1568), Danti (1573), Benedetti (1585), Kepler (1604), Fabricius (1611), Scheiner (1612), Schwenter (1636), Bettini (1645), Niceron (1646), Kircher (1646), Mersenne (1651), Schott (1657), Chérubin d'Orléans (1671), Zahn (1685). Plate 3.2 displays some of the explanations.

13. Line 145–6:

ويكون قوسا هذا الهلال من دائرتين متساويتين، لأن قوس الهلال المضيء من دائرتين متساويتين.

14. Line 54–7:

كل سطح مستو يقطع جسمًا مخروطًا ويكون ذلك السطح موازيا لقاعدة ذلك المخروط ... يكون شكله شبيها بشكل قاعدة المخروط. وقد تبيّن هذا المعنى في المقالة الأولى من كتاب المخروطات.

difference, for the plane parallel to the aperture is parallel to the plane of the circle bounding the Sun, whenever the aperture is opposite the Sun."[15]

Ibn al-Haytham's tolerance of approximate parallelism between the front side of the darkroom and the circle along which the cone, whose apex is the pinhole, is tangent to the Sun's sphere, suggests that the darkroom was a stationary device. This will be clearer after an examination of the darkroom's parallelism and size.

Parallelism. The rear face, which served as a projection plane, was plane and parallel to the front face. *On the Shape of the Eclipse* reads for instance: "For any circular aperture facing the luminous part of the Sun, behind which is arranged a projection plane parallel to the plane of the aperture (*saṭḥ muwāzī li-saṭḥ al-thaqb*)... light will appear crescent-shaped on the plane parallel to the aperture."[16] The wording *al-saṭḥ muwāzī li-saṭḥ al-thaqb* appears frequently (e.g., lines 34, 53, 69, 520, 531, 650, 652). Thus the two opposite faces of the darkroom were permanently arranged parallel to each other.

Aperture. The front face of the darkroom was possibly equipped with an aperture, which could vary in shape and size. *On the Shape of the Eclipse* makes clear that the aperture is either circular (*al-thaqb mustadīr*, lines 4, 25, 27, 36, 59, 72, 275, 276, 379, 529, 650) or square (*al-thaqb murabbaʿa*, lines 31–2), that it is either narrow (*al-thaqb ḍayyiq*, lines 4, 21, 37, 42, 520, 708) or wide (*al-thuqub al-waāsiʿa*, lines 28, 29, 43, 701, 706).

From the above, we conclude that (1) Ibn al-Haytham actually observed a partial eclipse of the Sun (2) through a simple stationary darkroom (3) whose aperture was located on the vertical front wall and (4) whose projection plane was the vertical rear wall of the darkroom. It follows that the device used by Ibn al-Haytham was a sim-

---

15. Line 89–93:
وسطح الثقب موازي لسطح الدائرة التي تماس عليها المخروط الأول محيط الشمس، أو ليس بينه، أعني سطح الثقب، وبين سطح الدائرة المحيطة بالشمس اختلاف محسوس؛ فالسطح الموازي الثقب مواز لسطح الدائرة المحيطة بالشمس ...

16. Line 530–5:
كل ثقب مستدير، إذا قوبل بالجزء المضيء من الشمس وكان وراء الثقب سطح مواز لسطح الثقب ... فإن الضوء يظهر على السطح الموازي للثقب هلاليًا.

ple, solid darkroom, similar to—or perhaps the same as—that used in the *Optics*, I, 3. There, Ibn al-Haytham mentionned that the darkroom was a properly caulked chamber, "allowing no light to come into the chamber except through the hole." The aperture was pierced in the upper part of the front wall: "The light should reach the facing wall through an aperture at the top of the wall of the dark chamber" (Sabra 1989, I: 22).

Finally, by taking together these conditions and the passage quoted in the *Movability* section, we obtain the following syllogism: if the darkroom has parallel vertical walls; if the plane where the cone is tangent to the Sun's sphere is parallel to the front face of the darkroom "with no appreciable difference;" then the solar eclipse was observed when the Sun was near the horizon.

3. Dimensions of the Camera Obscura

Depth of the Darkroom. To produce a bright image, the camera obsura must be large enough. Even though *On the Shape of the Eclipse* does not provide direct information on this, the depth—that is the focal distance of the darkroom—can be inferred from the manuscripts.

A fact which deserves attention is the consistency in size of the diagrams through all manuscripts. This is the sign that the scribes faithfully reproduced Ibn al-Haytham's autograph. Every manuscript provides representations of the eclipsed Sun in Diagrams 1 and 3, each of which provides five crescent-shaped images—an arrangement that supplies fifty images in total. Just consider the dimensions of the solar image:

F1	F3	B1	B3	P1	P3	O1	O3	L1	L3
23.4	26.2	28.2	28.2	22.2	29.8	32.0	31.1	38.0	42.5
23.4	26.2	28.2	28.2	22.2	29.8	32.0	31.1	38.0	42.5
28.6	24.4	28.2	18.4	34.6	35.7	33.7	30.2	28.1	19.7
27.4	24.4	20.8	18.4	34.6	35.7	34.6	30.2	29.6	19.6
29.8	24.4	23.3	18.4	34.6	35.7	34.6	30.2	28.1	19.4

The diameter of the solar disc ranges from 18.4 mm (MS B) to 42.5 mm (MS L), with a weighted average of 28.7 mm and a standard deviation of 6.0 mm. As these measures fit quite well with one another; in all likelihhod, the diameter of the solar disc in Ibn al-Haytham's autograph was in the interval 29 ± 6 mm.

Assuming now that the size of the image cast was of this order of magnitude, we apply the formula $\ell_1/r = \ell_0/R_0$, where r is the radius of the solar image, R_0 is the solar radius and ℓ_1 is the focal distance of the dark chamber (Fig. 3.6).

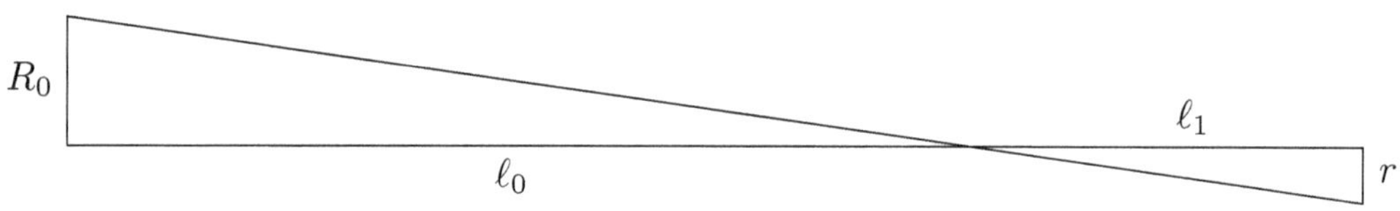

Fig. 3.6. Similar Triangles in the Darkroom

The known elements are the following: the Sun-Earth distance is $\ell_0 = 1$ AU (by definition) and the radius of the Sun is $R_0 = 0.004691$ AU. Therefore, a solar crescent with the radius $r = 14.5 \pm 3$ mm (half the diameter) yields a focal distance of the darkroom $\ell_1 = 3091$ mm—about 3 meters, which corresponds to the depth of an ordinary room.

Size of the aperture. At the beginning of the text, Ibn al-Haytham states that the aperture must be "narrow" to cast a crescent-shaped image of the eclipsed Sun.[17] This "narrow aperture" is word-for-word comparable to the "minute" or "tiny holes" referred to in Ibn al-Haytham's *Optics* I, 3. Thereafter, the size of the aperture is enlarged. The question therefore arises as to the minimum and maximum sizes of the aperture of the darkroom. Let us address them in turn.

1) It is relatively easy to determine the minimum size of the aperture, which is explicitly stated in the text: "The diameter of the aperture, by which the Sun is tested [or experimented], amounts to one grain of barley (*ʿaraḍa shaʿīra wāḥida*)" (line

17. Note that Ibn al-Haytham repeatedly says that the projection plane is "parallel *to the plane* of the aperture" (lines 34, 53, 69, 531). If the aperture had been wider, he would have been satisfied with saying that the projection plane was parallel *to the aperture*.

673).[18] This same configuration is echoed in the *Optics* where "a hole, one grain of barley in diameter" is used—the Latin version repeated "et dyameter foraminis est unius grani ordei" (Book 4, Prop. 3.68, Smith 2006: xx).

Assume first that Ibn al-Haytham specified the size of the aperture with no reference to any unit of measurement. *Hordeum vulgare* is cultivated from antiquity. Ethnobotanical and archeological specimens show no significant variation in size. This fixes the size of the grain of barley to about 2 mm (Fig. 3.7).

Fig. 3.7. Iron-Age Barley Grains (Bouby et al. 2011)

Barleycorn is also known as a unit of length by several Arab scholars, such as al-Farghānī (ca. 857, 1669: 74), al-Mas'ūdī (ca. 940, 1861: 183), al-Muqaddasī (ca. 985; de Goeje 1906: 65), *Kitāb al-ḥāwī* (XIth; Sauvaire 1886: 482), al-Idrīsī (1154; Jaubert 1840: 2), Abū al-Fidā' (ante 1321, 1848: 18), 'Alī al-Qūshjī (1457, 1652: 96). These authors report that six horsehairs (*sha'ar al-khīl*) are one barleycorn (*sha'īra*), six of which "laid belly-to-back" are one digit (*iṣba'a*). This fixes the average value of the grain thickness to about 3 mm.[19]

The two estimates are consistent with each other. They fix the minimum size of the pinhole to about 2–3 mm wide. As we shall see, this is all what is needed to state that Ibn al-Haytham investigated nearly stigmatic images.

---

18. Line 673:

فإذا كان قطر الثقب الذي يعتبر به الشمس عرض شعيرة واحدة ...

19. I retain 3 mm as an average value for this unit that fluctuated widely, and was most likely below the current value (Table, p. 222: Metric estimates; Sauvaire 1886; Hamilton 1946; Popper 1951–55; Hinz 1955; Mercier 1994). Those scholars who comment on Ibn al-Haytham's work and claim that one grain of barley amounts to 0.85 cm rely on an English unit (⅓ inch) which is a measure of the length, not the thickness, of the grain.

2) On the other hand, the maximum size of the aperture is deducible from a passage where the condition for obtaining a circular image is stated (lines 641–3). This is the inequality $R \geqslant 10\, R_0\, \ell_1/\ell_0$ which boils down to $R \geqslant 145$ mm. This means that Ibn al-Haytham experimented with apertures up to this diameter (for details, see below, Chapter 4.3, pp. 122–3).

6. Evaluating Ibn al-Haytham's Device

On the Shape of the Eclipse has two different facets. The first conveys knowledge about Ibn al-Haytham's optical theory and requires a historical assessment. The second provides information about what was really seen through the camera obscura, a fact that mainly depends on the physical characteristics of the instrument and calls for an assessment in the terms of the modern optics. This Section, which is critical to estimating the scientific reach of Ibn al-Haytham's treatise, is dedicated to establishing the optical performance of the darkroom that he used.

A large portion of this Section will be devoted to discussing the optical concept of stigmatism for two main reasons:

1. The size of the aperture just examined is the main factor for achieving a sharp image of the Sun. Could Ibn al-Haytham observe sharp images?

2. The theory of the pinhole images has been under discussion until recently. It is therefore important to base the present evaluation of Ibn al-Haytham's camera obscura on current knowledge.

In modern optics, stigmatism is the ability of an optical system to produce a sharp image of an object. If most optical systems suffer from chromatic and geometrical aberrations and only comply with approximated stigmatism, several systems are known to be rigorously stigmatic, such as the parabolic mirror, the spherical diopter and the stenope, i.e., camera obscura.

Optics provides two main approaches to define stigmatism. The first is formulated in purely geometrical terms; the second builds on a few notions of wave optics.

1. Stigmatism in Geometrical Optics

Let us recall the classical definition of stigmatism, which is based on the conjugation relationships initially proposed by Descartes on thin lenses. This notion was later developed by Gauss, Herschel and Abbe in the mid-nineteenth century.

Definition 1: An optical system is rigorously stigmatic — i.e., provides a sharp image — if all rays issued from one object-point A_0 go through one image-point A_1 once past the optical system (Fig. 3.8).

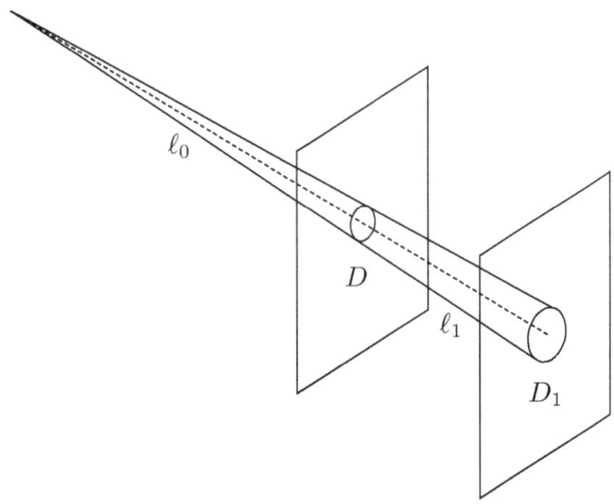

Fig. 3.8. Stigmatism and Beam Spreading

Turning now to the camera obscura, the image is rigorously stigmatic if the patch of light D_1 reduces to a point—a situation which precludes receiving light on the projection plane. The aperture must be a little enlarged, which results in a blurring of the image.

The diameter of the image is a linear function of the diameter of the aperture. For an object situated at a distance ℓ_0 from a darkroom of focal length ℓ_1 equipped with a pinhole of diameter D, the diameter of the light patch is

$$D_1 = \frac{D\,(\ell_0 + \ell_1)}{\ell_0}$$

As the Sun is at a distance much greater than the depth of the darkroom ℓ_1, sunrays are virtually parallel, hence $(\ell_0 + \ell_1)/\ell_0 \approx 1$. The depth of the darkroom is unnecessary data and the diameter of the light patch is equal to that of the aperture. Ibn al-Haytham's device deviated from rigorous stigmatism by the diameter of the aperture $D \approx D_1$, i.e., about 2–3 mm from data set out in Chapter 3.5, pp. 102–3.

2. *Stigmatism in Wave Optics*

Another method for judging the quality of the image is to investigate the phase differences between the waves terminating at the same image-point (Rayleigh 1879; Barakat 1965; Pérez 2000: 20–4). Consider visible light with wavelenght λ. An image is acceptable if peaks and valleys do not overlap. The $\lambda/4$ criterion is common practice to move away from destructive interferences arising when waves are in phase opposition at $\lambda/2$. This criterion was introduced early on: "So long as the difference of phase is less than a quarter of a period, the resultant cannot differ much from the maximum" (Rayleigh 1879: 262).

Definition 2: An optical system is rigorously stigmatic if the all rays issued from one object-point A_0 go through one image-point A_1 and if the outermost and axial optical paths differ by no more than $\lambda/4$ (Fig. 3.9).

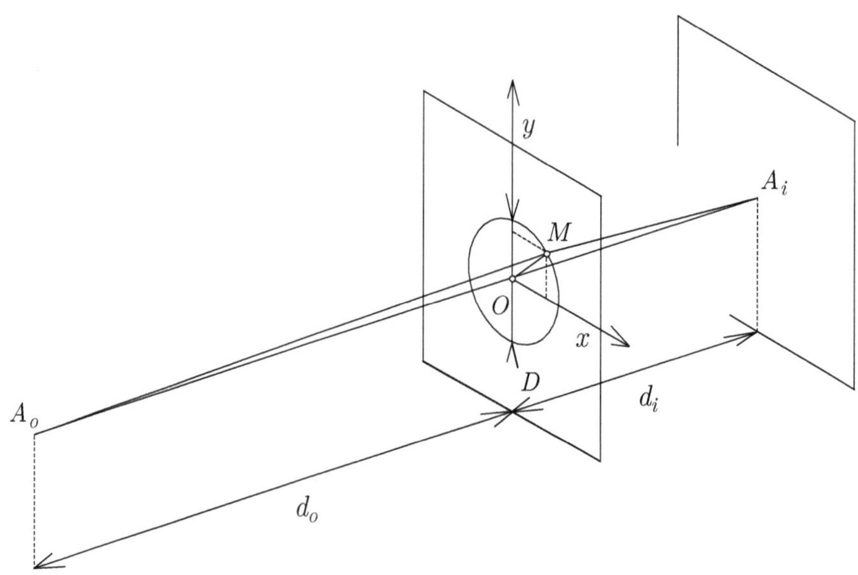

Fig. 3.9. Stigmatism and Optical Path Difference

The axial path goes through point O. The outer path goes through point M. Let us compare the lengths of the paths A_oOA_i and A_oMA_i.

(1) The axial path is simply $A_oOA_i = d_o + d_i$

(2) The outer path is made of two parts:

The first part is

$$A_oM = (d_o^2 + x^2 + y^2)^{1/2} \approx d_o + \frac{x^2 + y^2}{2d_o}$$

The second part is

$$MA_i \approx d_i + \frac{x^2 + y^2}{2d_i}$$

Therefore

$$A_oMA_i = d_o + d_i + \frac{x^2 + y^2}{2}\left(\frac{1}{d_o} + \frac{1}{d_i}\right)$$

(3) If we want $A_oOA_i \approx A_oMA_i$, we take diameter $D = 2\ OM$ so that

$$\frac{D^2}{8}\left(\frac{1}{d_o} + \frac{1}{d_i}\right) \leqslant \frac{\lambda}{4}$$

Given $\lambda = 555$ nm (average wavelength in photopic vision), $d_i = 3$ m (focal length of the darkroom) and $d_o = 1$ AU (Sun-Earth distance), we obtain

$$D_0 = \sqrt{\frac{2\lambda d_o d_i}{d_o + d_i}} = 1.926 \text{ mm}$$

It appears that this value is very similar to the diameter of the aperture computed separately above.[20]

20. Conversely, if one fixes the size of the aperture according to data provided in Section 5.3, pp. 102–3, d_i can be calculated in terms of D. The formula becomes

$$\left(\frac{1}{d_o} + \frac{1}{d_i}\right) = \frac{2\lambda}{D^2}$$

Then

$$\frac{1}{d_i} \leqslant \frac{2\lambda d_o}{D^2 d_o} - \frac{D^2}{D^2 d_o} = \frac{2\lambda d_o - D^2}{D^2 d_o}$$

And finally

$$d_i \geqslant \frac{D^2 d_o}{2\lambda d_o - D^2}$$

3. Diffraction

With an aperture 2 mm wide, we enter an area where diffraction may occur. Physical optics is then needed to evaluate the image quality.

The study of wavefront aberration conducted in the previous paragraph is not an investigation of the aberration of the image. Through a wide opening, the image of a distant object is large and fuzzy. When the diameter of the aperture reduces in size, resolution is improved up to a certain limit. If the diameter is further reduced, diffraction appears. There are two kinds of diffraction according to the value of the Fresnel number $F = D_0^2/\ell_1\lambda$:

— this is far-field diffraction if the diameter of the aperture is much smaller than the focal distance ($F \ll 1$);

— this is near-field diffraction if the diameter of the aperture is equal to or above the focal distance ($F \geqslant 1$).

Like many darkrooms that were subsequently built in the West, Ibn al-Haytham's chamber was subject to the far-field diffraction for $\ell_1 \gg \dfrac{D_0^2}{2\lambda}$.

If we consider a circular aperture, the patch of light will be in the form of the Airy disc. By contrast with geometrical optics, the radius of the image is now written as a hyperbolic function of the radius of the Airy disc $1.22\lambda\,\ell_1/D_0$, where λ denotes the wavelength of light, ℓ_1 is the focal length and D_0 is the diameter of the pinhole.

4. Sharpness of an Image

Geometrical and physical optics come under conflict on the subject of sharpness of an image. Geometrical optics encourages reducing the size of the aperture D_0 when diffraction recommends that this size is increased. In this context, one would expect the sharpest image to appear at the point of intersection of the two functions.

Several points are worth noting about this problem.

Given $D = 2$ mm, $d_o = 1$ AU $= 1.496 \times 10^{14}$ mm and $\lambda = 555$ nm, a sharp image would be produced with a focal length $d_i \geqslant 3\,600$ mm, a value that does not differ substantially from that computed separately from the size of the crescent-shaped images.

1. It is experimentally recognized that sharpness of the image appears when both geometric and diffractive blurrings are avoided, while continuously varying the diameter of the aperture (Fig. 3.10).

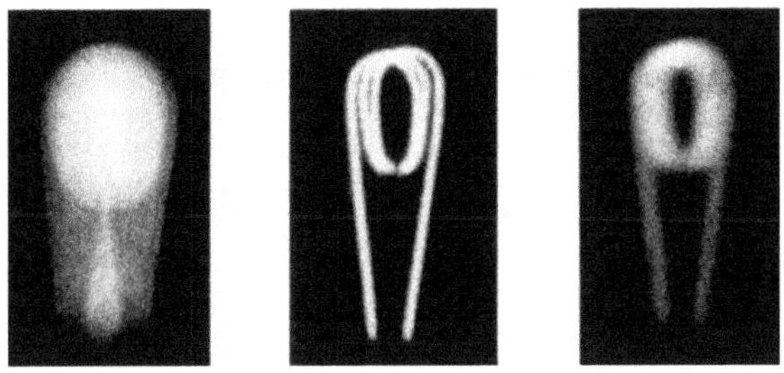

geometric blurring | sharpest image | diffractive blurring

Fig. 3.10. Image Sharpness and Blurring (Rüchardt 1958)

2. Following the research undertaken by Petzval (1857), Rayleigh (1891) and Abney (1895), many authors have provided equations for the minimum corresponding to the sharpest image. However, they disagree on the formulation of the law that sets the diameter of the aperture. The discrepancy between these formulas boils down to a mere difference of coefficient. Denoting ℓ_1 the focal length of the darkroom and D_0 the diameter of the aperture, all equations[21] are in the form

$$D_0 = \sqrt{k\,\lambda \ell_1} \qquad \text{with } 2.44_{\text{ Airy}} \leqslant k \leqslant 3.66_{\text{ Rayleigh}}$$

The issue is how to determine the correct k-value.

The previous sections have shown that geometrical optics writes the radius of the image as a linear function of the radius of the aperture, while physical optics writes the radius of the image as a hyperbolic function of the radius of the Airy disc. As linear and hyperbolic functions rise in opposite directions, they intersect at one point. Therefore one would expect the sharpest image to appear at the point where the two functions intersect. Experimental approach of the subject has shown that reality is,

21. Note, however, that these equations hold at infinity only. When the light source is at some finite distance ℓ_0, the equation needs to be rewritten as $D_0 = \sqrt{k\lambda\,\ell_0\ell_1/(\ell_0+\ell_1)}$. One can check that this formula satisfies $\ell_0\ell_1/(\ell_0+\ell_1) \approx \ell_1$ when $\ell_0 \to \infty$.

however, a little different. This is seen when experimental data of both functions are plotted in an image diameter vs. pinhole diameter diagram.

1. Experimental optics has shown that real data providing the minimum diameter of the image regularly fall in the region of, rather than at, the exact point of intersection of these functions (Young 1971–1989; Carlsson 2004). The intersection is in a region where neither geometrical and far-field approximation applies. Strictly speaking, these functions are *undefined* in the region where the minimum occurs.

2. Recent research has shown that image sharpness is subjected to perceptual factors. Two factors come into play in the evaluation of the image quality: resolution and contrast. In the equation of D_0, resolution provides a value near $k = 3.66_{\text{Rayleigh}}$ while contrast gives a value near $k = 2.44_{\text{Airy}}$. The analysis of image has been refined by the use of the PSF point spread function and its Fourier transform, known as the MTF modulation transfer function (Sayanagi 1967; Mielenz 1999). When the MTF is plotted against spatial frequency $\xi = 1/\sqrt{\lambda \ell_1}$ for different radii R of the aperture (in $\sqrt{\lambda \ell_1}$ units), the intersecting of the curves confirms that several k-values are admittable. The curve $R = 0.78$ reaches high values for low or mid-spatial frequency and provides higher contrast, while the curve $R = 1.0$ keeps significant values for high spatial frequency and offers higher resolution. The key point here is that recent research on image quality has revealed that human vision prefers high contrast to high resolution (Carlsson 2004; Nyman et al. 2006; Emmel 2011). Therefore, the sharpest image is obtained with an aperture of radius $0.78\sqrt{\lambda \ell_1}$. If we consent to sum up this knowledge in one number, Airy's value $k = 4\,(0.78)^2$ should be preferred, thus:

$$D_0 = \sqrt{2.44\,\lambda \ell_1}$$

5. Optimal Size of the Aperture

Given $\lambda = 555$ nm (mean wavelength in photopic vision), $\ell_1 = 3091$ mm (about 5 ½ *dhirā'*, the focal length of Ibn al-Haytham's darkroom), the size of the aperture providing the sharpest image is $D_0 = 2.015$ mm.

This value is not very different from the minimum size of the aperture stated in the text: "The diameter of the aperture, by which the Sun is experimented, amounts to one grain of barley (*sha'īra wāḥida*)" (line 673). As seen before, historical information by the Arab scholars (al-Farghānī, al-Mas'ūdī, al-Muqaddasī, etc.) and archeobotanical data coincide in fixing the thickness of one grain of barley to about 2–3 mm (Section 5.3, pp. 102–3).

In conclusion, although the modern concept of stigmatism was of course unknown to Ibn al-Haytham, he seems to have experimented with small apertures capable of producing near stigmatic images:

> "A crescent-shaped image of the light of the Sun can be seen at the time of the eclipse, if its light has passed through a narrow circular aperture and has reached a plane opposite the aperture ... The image that results from the light of the Sun at the time of its eclipse, only appears such <i.e., crescent-shaped> if the aperture is narrow and up to some extent of capacity" (lines 3–4, 20–1).

Most interestingly, we note that throughout *On the Shape of the Eclipse* the word *ṣūra* is used consistently in passages where the aperture is narrow.[22] This result opposes Sabra's opinion that "There is no description in I.H.'s *Optics* of a picture obtained by means of a narrow opening, nor is such a description to be found in any of his extant writings" (Sabra 1989, I: li). As we have seen, there are solid grounds to consider that Ibn al-Haytham experimented with both large and pinhole apertures and possessed the concept of "stigmatic image" in the modern sense of the word.

22. Lines 2, 3, 11, 20, 22, 23, 39. The only exception is when the word *ṣūra* refers to the outline of the crescent-shaped image: "the *image* of the concavity that appears in this light..." (line 439).

Chapter 4
Ibn al-Haytham's Optical Analysis

This chapter aims to assess the findings of *On the Shape of the Eclipse* by explaining its reasoning, and by comparing it, firstly to objective data, and secondly to the results of modern optics. This should provide a test of robustness of the results obtained in this work.

1. Conditions for an Image to be Seen in the Darkroom

Ibn al-Haytham research may only be assessed by relating it to the physical situation it intends to describe and explain. There are several conditions to be met for an image to be seen in the camera obscura:

— The object must be luminous enough;
— The room must be dark enough;
— The room must be wide enough;
— The hole must be narrow enough to cast a sharp image;
— But not too narrow to avoid diffraction.

As much is said in *On the Shape of the Eclipse* about light passing through wide apertures, I first consider the image produced through a narrow aperture. In this situation, the first relevant observation deals with the image inversion. Then I consider stigmatism (image sharpness) in geometrical optics and wave optics. This will lead us to determine the optimal size of the pinhole of Ibn al-Haytham's camera obscura, all other things being known. It is only after this step that the geometry of the image passed through wide apertures will be discussed.

D. Raynaud, *A Critical Edition of Ibn al-Haytham's On the Shape of the Eclipse*,
Sources and Studies in the History of Mathematics and Physical Sciences,
DOI 10.1007/978-3-319-47991-0_4

2. Image Inversion

As surprising as it may seem, I will not start this section with a presentation of Ibn al-Haytham's tangible results on image inversion, but rather with a brief reminder of the views that the historians of science have expressed on his contribution. This is, in fact, the only way to correctly assess his understanding of the darkroom. The historical approach is particularly welcomed to discover why Ibn al-Haytham's knowledge of the formation of the image is not clearly expressed in literature.

In an article of the *Encyclopaedia of Islam*, Wiedemann claimed that

> "Kamāl al-Dīn al-Fārisī was a scholar equal in calibre to Ibn al-Haitham and, indeed, perhaps surpassed him in originality. Quṭb Dīn al-Shirāzī had called his attention to the latter's *Optics*, which he procured and wrote an excellent commentary upon. He added a series of brilliant treatises to it. These deal more particularly with the refractions and reflections of a sphere, the rainbow, the halo, *camera obscura*, etc. As to the the latter it should be noted that the first scholar whom we know to have used the *camera obscura* was Ibn al-Haitham [...] Kamāl al-Dīn gave a more perfect theory and tested it by brilliant experiments. He first made the orifice very small <a> and placed opposite it a surface half red and green. He then showed how one got the sharper images the smaller the opening <b> and that the images were independent of the shape of the orifice <c>. The large the opening the less these principles applied <d>. It was to be noted that the images were reversed <e>. With this apparatus Kamāl al-Dīn observed on the wall the clouds and their movements as well as a bird flying past <f>. The movements in the image are in the contrary direction to real life [...]" (Wiedemann 1913–36: 704).

Characters <a-e> are by no means specific to Kamāl al-Dīn al-Fārisī's research. They in fact date back to previous works, in particular *On the Shape of the Eclispe* as can be seen from a comparison with Ibn al-Haytham's text.

<a> lines 4, 21, 37, 43, 520, 673, 708.

<b> lines 2–3, 11, 20–23, 39.

<c> lines 3–5, 21.

<d> lines 21–23.

<e> lines 141, 145, also Mo Zi, cited p. 24.

<f> also Mo Zi (Needham 1962: 82).

In his preface to Ibn al-Haytham's *Optics*, Sabra expresses a view close to Wiedemann's appreciation. Even if Ibn al-Haytham is said to have seen "a reversed crescent image cast on the screen," all attention is drawn to the fact that

> "Kamāl al-Dīn, by contrast, reported observations of reversed coloured images (*ṣuwar*) of clouds and birds on an unlighted white surface place behind a narrow opening" (Sabra 1989, II: l–li).

This is correct, but the attribution of the concept of image to Kamāl al-Dīn has the effect to limiting Ibn al-Haytham's contribution to the study of fuzzy patches of light, which is not true.

Thereafter, the side-by-side presentation of Ibn al-Haytham's and Kamāl al-Dīn's work on the image inversion has become something of a tradition. One may wonder whether this presentation—which is detrimental to Ibn al-Haytham—did not contribute to the confusion of the situation.

Let us look now at Ibn al-Haytham's contribution.

Ibn al-Haytham describes the image inversion in a short passage in *On the Shape of the Eclipse*. During the partial solar eclipse, he says, "the convexity of the arc [cast on the projection plane] will be *opposite* (*ḍidd*) the convexity of the self-luminous crescent, which is part of the Sun" (lines 141–2).

Things are even clearer when he declares that the crescent-shaped image of the sunlight consists of "two arcs similar to those of the light crescent—*though contrary in position*" (*wa-yakūn waḍ'ahu mukhālif^{an} li-waḍ'ihi*, line 145).

Furthermore, it has to be noted that this is exactly how the crescent-shaped images are arranged in all diagrams (Fig. 4.1).

Ibn al-Haytham provides a very simple explanation for this phenomenon. He considers a single lightray moving freely around one point of the aperture taken as a pivot, say point Ḥ, and demonstrates that the movement of the endpoint of this line on the projected crescent is coupled to that of the other endpoint on the solar crescent: "The line that comes point Ǧ to point Ḥ, ends at point Š. If the apex of the light cone moves from point Ǧ to point B, the endpoint of the line, which is point Š, will move along the perimeter of the circle whose center is point K" (lines 395–8).

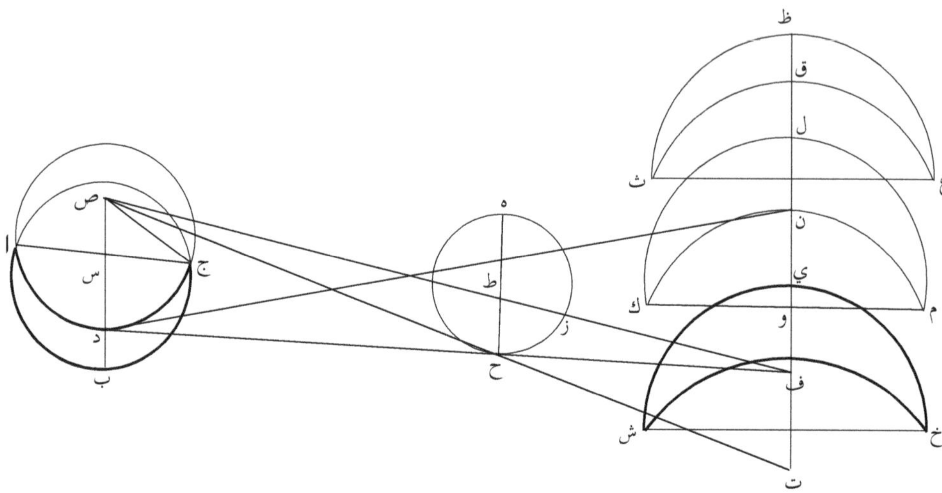

Fig. 4.1. Image Inversion as on MS P

A textual analysis shows that Ibn al-Haytham succeeded in using a pinhole camera and in explaining the image inversion, even if he did so without being verbose on the subject.

The side-by-side presentation of Ibn al-Haytham's own contribution with Wiedemann and Sabra's appreciations of Kamāl al-Dīn's research shows that the achievements granted to the latter are not of his own making. These comments cloud the fact that Ibn al-Haytham successfully described, and geometrically explained, the inversion of the image in the camera obscura.

3. Outline of the Demonstration

This Section outlines the structure of Ibn al-Haytham's reasoning in *On the Shape of the Eclipse*, and highlights the relations that constitute the milestones of his line of reasoning. Before recalling his proof, let us recall that a proportion A : B :: C : D will be denoted in modern terms as $A/B = C/D$. As the same letters L N Y F are used to denote the top points and centers of the arcs in the original diagrams, let us rename the centers L′ N′ Y′ F′. Finally, as the circumlocution "the line which is between point Ṣ and the line going through points T Ṭ, and ends at line BṢ" is a little cumbersome; let us simply call X the intersection point of lines BṢ and TṬ.

1. Mathematical Relationships

Having set forth his observations and the principles underpinning his demonstration, Ibn al-Haytham prepares the elements to compare the distance NF between two arcs, which depends on the size of the aperture, and the radius FT of each arc, which is constant for a given darkroom. (Note in advance that the image is blurring when NF > FT, while is sharpening when NF < FT.)

This comparison is done on Diagram 1 that aligns the eclipsed Sun (on the left, the Sun below) with its image (on the right). This figure depicts how the partially eclipsed Sun ABǦD casts through the aperture HṬḤ a composite image on the projection plane ẒT. This image is formed by the overlapping of image ŠYḪF obtained through Ḥ, image KLMN obtained through Ṭ, image ṮẒGQ obtained through H, as well as any other image produced by some point of the aperture (Fig. 4.2).

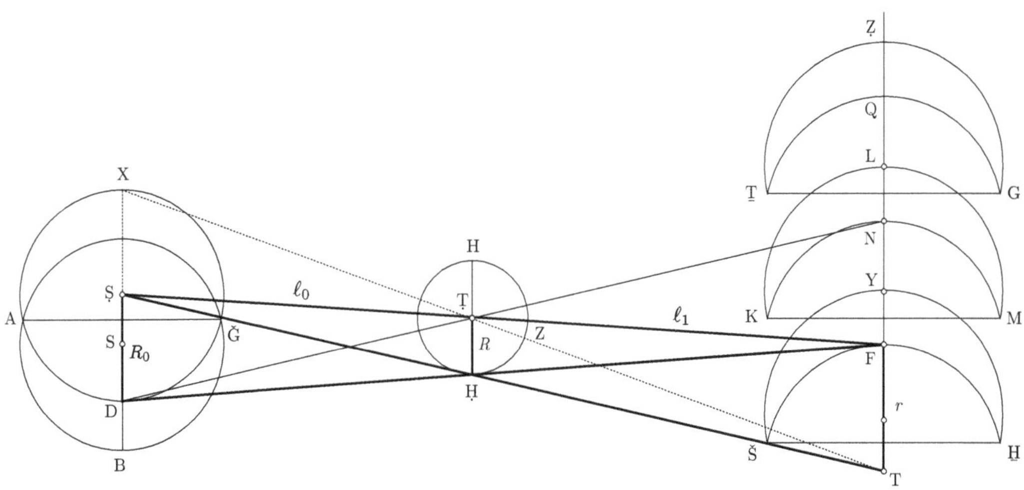

Fig. 4.2. Proportionality when NF = FT

With this done, Ibn al-Haytham establishes two kinds of relationships which constantly intervene in his demonstration.

- The first is proportionality in opposite triangles (e.g., lines 229–30):

$$\frac{\text{FT}}{\text{ṢD}} = \frac{\text{FḤ}}{\text{ḤD}} \quad \text{or equivalently,} \quad \frac{\text{FT}}{\text{ṢD}} = \frac{\text{FṬ}}{\text{ṬṢ}}, \quad \text{etc.}$$

This is an essential (permanent) relationship. With R_0 the radius of the Sun, ℓ_0 the Sun-Earth distance (Ibn al-Haytham uses the values found in the *Almagest*, i.e., $5\frac{1}{2}$ and 1,210 Earth radii respectively), r the radius of the image, and ℓ_1 the focal distance of the chamber, this boils down to

$$r/R_0 = \ell_1/\ell_0$$

- The second is proportionality in similar triangles. In particular (e.g., line 195), Ibn al-Haytham investigates the condition under which:

$$\frac{\text{ṬḤ}}{\text{ṢD}} = \frac{\text{FṬ}}{\text{FṢ}} \quad \text{or equivalently,} \quad \frac{\text{ṬḤ}}{\text{ṢD}} = \frac{\text{FḤ}}{\text{FD}}, \quad \text{etc.}$$

However, as the top point F of arc ŠFH̱ is not necessarily identical to the center F′ of arc KNM (Fig. 4.3), this relationship is one accidental (occasional), which just comes as a case. With the use of the same notations—with the exception of R, the radius of the aperture—this is written as $R/R_0 = \ell_1/(\ell_0 + \ell_1)$ (compare the respective places of points F and F′ in Figs. 4.2 to 4.4).

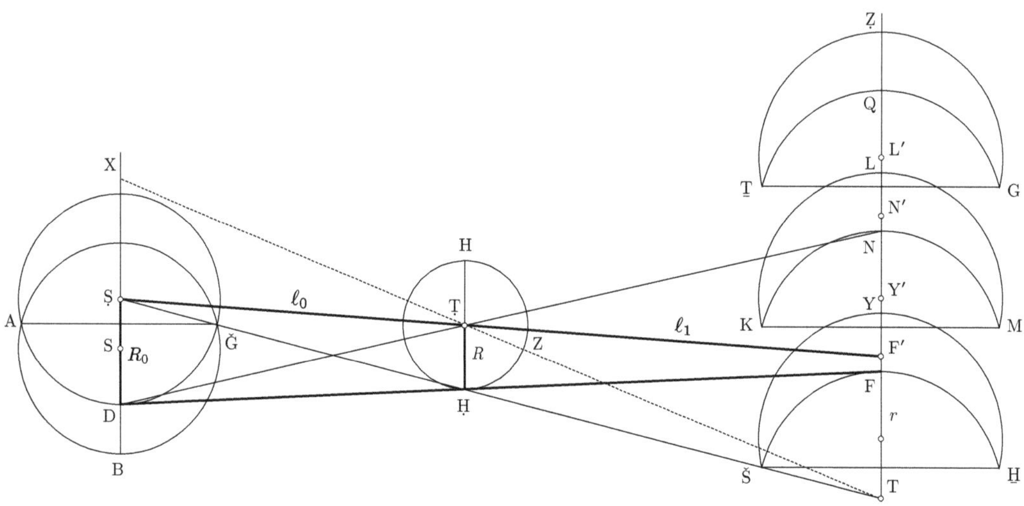

Fig. 4.3. Proportionality when NF > FT

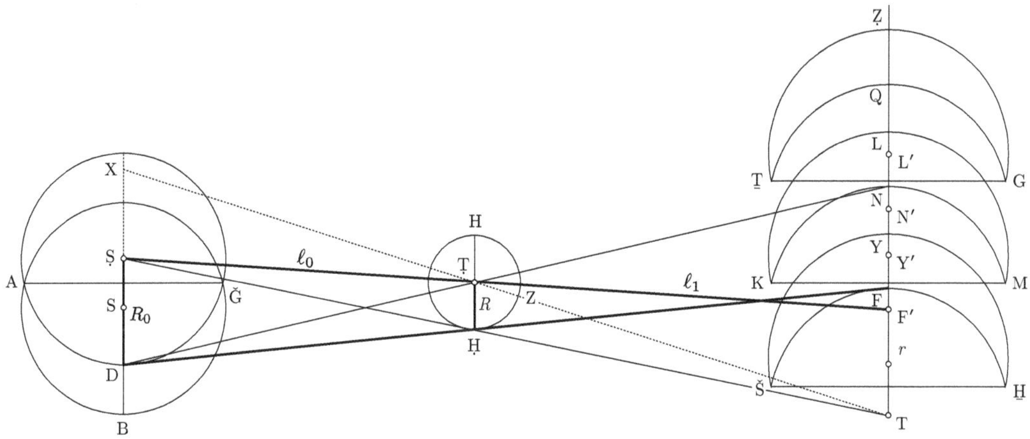

Fig. 4.4. Proportionality when NF < FT

The comparison of the ratio of diameters to the ratio of distances, and their consequences on the spacing NF = F′T of two adjacent crescents, are expressed in these terms in *On the Shape of the Eclipse*:

"If the ratio of (the semi-diameter of the aperture) to (the semi-diameter of the Sun) is smaller than the ratio of (the distance between the aperture and the plane parallel to the aperture) to (the distance between this plane and the Sun), then line F′T, which is the distance between the centers of the two arcs, will be smaller than the semi-diameter of the arc KNM. <Conversely> if the ratio of (the semi-diameter of the aperture) to (the semi-diameter of the Sun) is greater than the ratio of (distance) to (distance), then line F′T will be greater than the semi-diameter of the circle KNM" (lines 203–9).

The covariation of line ṢX with the distance F′T, which is depicted in Figs. 4.2 to 4.4, is stated in the text through another proportion in direct similar triangles, whose term ṢX is variable (e.g., lines 187–9):

$$\frac{\text{ṬḤ}}{\text{ṢX}} = \frac{\text{TḤ}}{\text{TṢ}}$$

To recapitulate: when the aperture shrinks, or when the focal distance of the dark room increases, the ratio R/R_0 is smaller than $\ell_1/(\ell_0 + \ell_1)$, ṢX reduces, and thus NF is smaller than FT. On the other hand, when the aperture widens, or when the focal distance reduces, the ratio R/R_0 is greater than $\ell_1/(\ell_0 + \ell_1)$, ṢX increases, and thus NF is greater than FT.

Ibn al-Haytham then describes the main properties of the image cast on the projection plane, once the image of the solar eclipse reduces to its central (KLMN) and extremal (ŠYḪF, ṮẒGQ) components. He proves the equality of the radii of the lower, middle and upper crescents (lines 184, 234).[23]

Consider the proportion (Fig. 4.4):

$$\frac{\mathrm{FT}}{\mathrm{ṢD}} = \frac{\ell_1}{\ell_0}$$

If straight line DṬ is extended to point N, then

$$\frac{\mathrm{NF'}}{\mathrm{ṢD}} = \frac{\ell_1}{\ell_0}$$

These relationships imply

$$\mathrm{FT} = \mathrm{NF'}$$

Ibn al-Haytham proceeds in a similar manner with the equality of the radii of the lower and upper arcs of any crescent (lines 233–5):

$$\mathrm{LY'} = \mathrm{NF'}\ (= \mathrm{YR'} = \mathrm{ẒL'} = \mathrm{QN'} = \mathrm{FT})$$

The same reasoning is applied to lateral magnitudes (lines 240–4):

$$\frac{\mathrm{ŠḪ}}{\mathrm{AĞ}} = \frac{\mathrm{FḤ}}{\mathrm{ḤD}}, \quad \frac{\mathrm{KM}}{\mathrm{AĞ}} = \frac{\mathrm{FḤ}}{\mathrm{ḤD}}, \quad \text{etc.}$$

From which it follows that

$$\mathrm{ṮG} = \mathrm{KM} = \mathrm{ŠḪ}$$

---

23. Condition F = F′ is unecessary here, so I skip the beginning of the proof, which starts as:

$$\frac{\mathrm{TḤ}}{\mathrm{TṢ}} = \frac{\mathrm{ṬḤ}}{\mathrm{ṢD}} \text{ and } \frac{\mathrm{TḤ}}{\mathrm{TṢ}} = \frac{\mathrm{ṬḤ}}{\mathrm{ṢX}}$$

imply ṢX = ṢD. Inserting an intervening term ($a : c = a : b \times b : c$),

$$\frac{\mathrm{FT}}{\mathrm{ṢD}} = \frac{\mathrm{FT}}{\mathrm{ṬḤ}} \times \frac{\mathrm{ṬḤ}}{\mathrm{ṢD}} = \frac{\mathrm{TṢ}}{\mathrm{ṢḤ}} \times \frac{\mathrm{ḤT}}{\mathrm{TṢ}}$$

Deleting TṢ, this leads to, etc.

Ibn al-Haytham then distinguishes between the convex and concave faces of the crescent, which require separate analyses.

The analysis of the convex face (lines 238–355) raises no special difficulty. Ibn al-Haytham introduces a lemma—"Whatever two equal circles, any straight line drawn parallel to the line joining the two centers will be equal to the line joining the two centers" (lines 281–3; Diagram 2)—as a basis for the continued demonstration.

That done, he demonstrates that the convex face of the solar crescent is imaged as a circular arc. He begins by highlighting

$$\text{KŠ} \parallel \text{NF} \qquad \text{KŠ} = \text{NF}$$

These relationships hold true of any point taken on two adjacent crescents. Consider a series of circular lights with their centers situated on arc KLM. Their perimeters reach arc ṮẒG. Since these lights overlap with one another, their perimeters form a common circular arc (lines 110–2). This property is otherwise evidenced by the Archimedean analysis (lines 356–438) discussed earlier in Chapter 3.2. This mechanical means—a simple straight edge extended between the Sun, the pinhole and the image—leads to the same conclusion: "the perimeter of light ... is a proper circular line, even if the curve does not achieve a full surround" (lines 437–8).

The analysis of the concave face of the crescent (lines 439–649) is far more advanced. Consider a series of circular lights with their centers situated on arc KNM. Each of these circles cuts arc ŠFH̱ (except one circle, which is that with center N). The region below the concave face of the crescent is lit, and the perimeter of light does not coincide with the circular arc ŠFH̱. This is a special curve tangent to all of the circles drawn from arc KNM.[24] This curve is studied by varying the size of the aperture and the focal length of the dark chamber (lines 472–529).

Ibn al-Haytham subsequently formulates the conditions for light to appear crescent-shaped or circular. The mathematical relationships set forth in earlier sections thus intervene as conditions to decide upon the shape of the image cast on the projection plane.

24. The text makes explicit that "the concave arc is not of perfect circularity, though it looks round to the sense" (lines 488–9).

2. The Crescent-Shaped Image

In Section 4.3 (*A Condition for Light to Appear Crescent-Shaped or Circular*), Ibn al-Haytham demonstrates that, "the closer the aperture to the plane parallel to it, the less sharp the concavity" (line 523). Indeed, when the aperture is close to the projection plane, "the line between the centers of the two arcs, is greater than the semi-radius of each arc" (lines 510–1). On this basis, he states: "If ṬḤ/ṢD $\leqslant \ell_1/\ell_0$, then the light appears crescent-shaped on the plane parallel to the aperture" (lines 531–5). This condition may be restated as

$$R \leqslant R_0\, \ell_1/\ell_0$$

Sabra interprets this relationship—written $m_{\mathrm{A}}/m_{\mathrm{S}} \leqslant d_{\mathrm{A}}/d_{\mathrm{S}}$ in his article—as "a condition for obtaining a *distinct image* of an object through a circular aperture" (1972: 196, italics mine). However, as Ibn al-Haytham only speaks in terms of a "crescent-shaped figure," this is not necessarily a distinct image.

The meaning of the rule ṬḤ/ṢD $\leqslant \ell_1/\ell_0$ should then be clarified. Given $R_0 = 0.004691$ AU, $\ell_0 = 1$ AU and $\ell_1 = 3091$ mm, the image is crescent-shaped if the radius of the aperture is $R \leqslant 14.5$ mm. It is clear that this radius is much too wide to form a "distinct image." A sharp image of the Sun only appears when the size of the aperture "amounts to one grain of barley" (*ʿaraḍa shaʿīra wāḥida*), as is stated in the text. As we are going to see further on, this condition is equivalent to the case $R \leqslant r$ discussed below in Chapter 4.4, pp. 133–4.

Consequently, $R \leqslant R_0\, \ell_1/\ell_0$ reflects a much looser condition, encompassing any crescent-shaped image from the sharp image ($R = 0$) to the limit where the cusp appears ($R = r$).

3. The Circular Image

In Section 4.4 (*Geometric Demonstration*), Ibn al-Haytham formulates a similar condition for the image to be circular: "If $\ell_1/(\ell_0 + \ell_1) \leqslant$ ṬḤ/(10 ṢD) ... then light will appear circular" (lines 536–41). Factor 10, which is involved in this reasoning, is not justified in the text. This is empirical choice: the light patch is almost circular if the cavity below arc ŠFH̱ is small enough not to be detected by the sense.

The previous condition may be written $\ell_1/\ell_0 \leqslant R/(10\,R_0)$ or, equivalently,

$$R \geqslant 10\,R_0\,\ell_1/\ell_0$$

Once ℓ_1 is picked in Chapter 3.5, pp. 101–2, all terms of $R \geqslant 10\,R_0\,\ell_1/\ell_0$ are known. It can be shown that the radius of the aperture at which the image is circular is 145 mm. Ibn al-Haytham then proceeds as follows on Diagram 4 (Fig. 4.5), which isolates two crescents to estimate the depth of cavity GR of the image as a function of the distance between the centers of the two crescents ŠFH̱ and KNM. The result is given as a ratio of GR over the diameter of arc ŠFH̱ (lines 613–4).

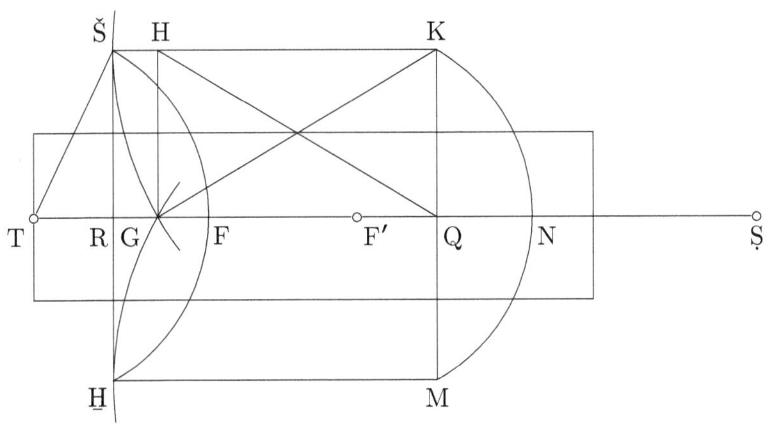

Fig. 4.5. The Determination of Line GR

Several relationships holds true by combining Fig. 4.4 with Fig. 4.5. They can be written in modern terms as:

$$\frac{\mathrm{FT}}{\mathrm{\d{S}D}} = \frac{\ell_1}{\ell_0} \qquad \frac{\mathrm{TF'}}{\mathrm{\d{S}X}} = \frac{\ell_1}{\ell_0} \qquad \frac{\mathrm{\d{T}\d{H}}}{\mathrm{\d{S}X}} = \frac{\ell_1}{\ell_0 + \ell_1} \tag{1}$$

[We are given the condition]

$$\frac{\mathrm{\d{T}\d{H}}}{10\ \mathrm{\d{S}D}} = \frac{\ell_1}{\ell_0 + \ell_1} \tag{2}$$

[(1c) and (2) imply] $\mathrm{\d{S}X} = 10\ \mathrm{\d{S}D}$ (3)

[(1a-b) and (3) imply] $\mathrm{TF'} = 10\ \mathrm{FT} = 10\ \mathrm{NF'}$

[Inserting the value of (3) in (1b) gives]

$$\frac{\mathrm{TF'}}{10\ \mathrm{\d{S}D}} = \frac{\ell_1}{\ell_0} \tag{4}$$

Also it is established that

$$\frac{\mathrm{NF'}}{\mathrm{\d{S}D}} = \frac{\ell_1}{\ell_0} \tag{5}$$

[Therefore (4) and (5) together imply]

$$\frac{\mathrm{TF'}}{10\ \mathrm{\d{S}D}} = \frac{\mathrm{NF'}}{\mathrm{\d{S}D}}$$

Which, alternating, is equivalent to

$$\frac{10\ \mathrm{\d{S}D}}{\mathrm{\d{S}D}} = \frac{\mathrm{TF'}}{\mathrm{NF'}} \text{ (lines 583–5).} \tag{6}$$

Choose point G on line TQ so that KG = KŠ = QR.

[Squaring KG = QR provides]

$\mathrm{KG}^2 = \mathrm{QR}^2$.

On straight line TQ, draw QṢ = QR.

[Line GṢ is the sum of GQ and QṢ, therefore]

$(\mathrm{G\d{S}} \times \mathrm{GR}) + \mathrm{QG}^2 = \mathrm{QR}^2$

[But KG = QR, thus]

$(\mathrm{G\d{S}} \times \mathrm{GR}) + \mathrm{QG}^2 = \mathrm{KG}^2$

As $\mathrm{KG}^2 = \mathrm{KQ}^2 + \mathrm{QG}^2$, these two relationships imply

$$\mathrm{KQ}^2 = \mathrm{G\d{S}} \times \mathrm{GR} \tag{7}$$

[The identity KQ = ŠR is equivalent to]

$$\mathrm{KQ}^2 = \mathrm{\check{S}R}^2. \tag{8}$$

[Therefore (7) and (8) gives]

$$\mathrm{G\d{S}} \times \mathrm{GR} = \mathrm{\check{S}R}^2 \text{ (line 598).} \tag{9}$$

[Line ŠR < FT because arc ŠF is less than $\frac{1}{4}$ circle. QR = 20 FT, thus]

$$\text{ŠR} < \tfrac{1}{10}\,\text{QR} \tag{10}$$

Combining (9) and (10) provides

$$\text{GṢ} \times \text{GR} < \tfrac{1}{100}\,\text{QR}^2 \text{ (line 606).}$$

Raise GH perpendicular to RQ, and draw the diagonals QH = KG = QR. [As points R H Ṣ are on the semi-circle of radius QR, line HG is the geometric mean of lines GR and GṢ. The relationship $\text{HG}^2 = \text{GR} \times \text{GṢ}$ characterizing the geometric mean may be written as]

$$\frac{\text{GR}}{\text{HG}} = \frac{\text{HG}}{\text{GṢ}} \tag{11}$$

[Assuming that the only known magnitude is ṢR, finding GR is an indeterminate problem. Ibn al-Haytham thus introduces an estimate of HG]

$$\text{HG} \approx \tfrac{1}{20}\,\text{GṢ}, \tag{12}$$

[By combining (11) and (12), we find

$$\text{GR} = \tfrac{1}{400}\,\text{GṢ} \tag{13}$$

The equality ṢR = GṢ + GR implies

$$\text{ṢR} = \tfrac{401}{400}\,\text{GṢ}$$

Thus ṢR ≈ GṢ. As a result (13) may be written as]

$$\text{GR} \approx \tfrac{1}{400}\,\text{ṢR} \tag{14}$$

[As ṢR = 20 FT, (14) is equivalent to]

$$\text{GR} \approx \tfrac{1}{20}\,\text{FT}$$

[And finally]

$$\text{GR} \approx \tfrac{1}{40}\,\text{diameter (ŠFH̱) (lines 613–4)}$$

As this ratio is "very small," the circular segment ŠFḪR is virtually full of light: "Nothing remains from the concavity of arc ŠFḪ, except a small part that is not perceived by the sense" (lines 628–9). Moreover, if the equal sign of the starting equation is replaced by a greater than or equal to sign, the light will be even rounder for line GR will be smaller and thus becomes unnoticeable (lines 631–2).

As we can see, the core of this demonstration consists in determining the magnitude of line GR from the geometric mean $\mathrm{HG}^2 = \mathrm{GR} \times \mathrm{GṢ}$. However, Ibn al-Haytham does not inform us why he opted to solve the problem through an approximation of HG—"half tenth of GṢ, rounded off" (lines 610–1). Whatever it might be, this approach is quite surprising on his part because the exact magnitude of the line HG = ŠR is knownable.

For the sake of clarity, I provide this addendum to the demonstration in modern terms.

The equation of the lunar circle ŠF writes $(x + \mathrm{TR})^2 + y^2 = \mathrm{FT}$. The equation of the solar circle ŠY is $(x - \mathrm{TR})^2 + y^2 = \mathrm{FT}$. Point Š belong to both circles; thus it fulfills both these equations (Fig. 4.6).

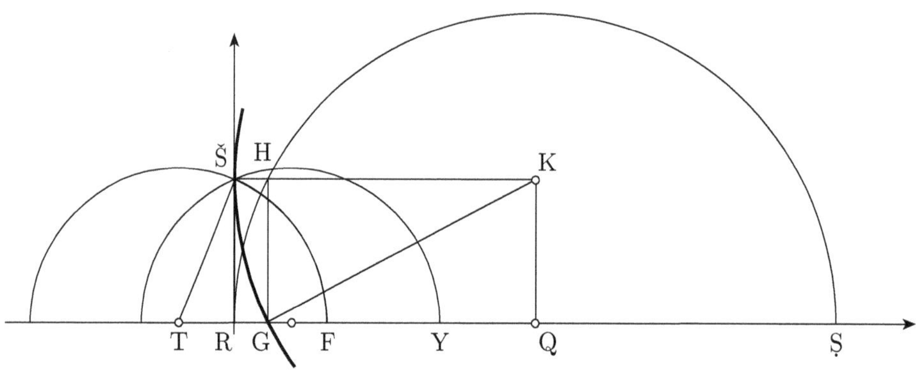

Fig. 4.6. The Determination of Ratio GR/ṢR

By symmetry $x = 0$, therefore $\mathrm{TR}^2 + y^2 = \mathrm{FT}$. Consequently,[25]

$$\mathrm{HG} = \check{\mathrm{S}}\mathrm{R} = \sqrt{\mathrm{FT} - \mathrm{TR}^2} \qquad (15)$$

25. A simplification could be considered. Line TR, which is half the distance between the two centers, depends on the magnitude g of the eclipse, thus $\mathrm{TR} = (1 - g)\,\mathrm{FT}$.

Why is it that Ibn al-Haytham calls the perimeter of light circle ŠG, and nevertheless considers the geometric mean on circle RHṢ? A downwards translation of circle ŠG ($y' = y - \mathrm{KQ}$), then a flipping around axis TṢ ($y'' = -y' = \mathrm{KQ} - y$) shows that line HG is in the same position vis-à-vis both circles. Therefore the statements: "HG is the geometric mean of ŠH and 2 ŠK – ŠH on circle ŠG" and "HG is the geometric mean of GR and GṢ on circle RHṢ" are equivalent.

As HG = ŠR, we plug (15) into the equation of the geometric mean, and we find, after several steps of simplification, the sagitta[26] $\mathrm{GR} = 10\,\mathrm{FT} - \sqrt{100\,\mathrm{FT}^2 - \mathrm{HG}^2}$, which may be written as $\mathrm{GR} = 10\,\mathrm{FT} - \sqrt{100\,\mathrm{FT}^2 - \mathrm{FT} + \mathrm{TR}^2}$. Line TR, which is half the distance between the two centers, is fixed by $\mathrm{TR} = (1 - g)\,\mathrm{FT}$.

With $g = 3/4$, we get $\mathrm{TR} = \frac{1}{4}\mathrm{FT}$. Thus,

$$\mathrm{GR} = 10\,\mathrm{FT} - \sqrt{100\,\mathrm{FT}^2 - \mathrm{FT} + \tfrac{1}{16}\,\mathrm{FT}^2} = 10\,\mathrm{FT} - \sqrt{\tfrac{1601}{16}\,\mathrm{FT}^2 - \mathrm{FT}}$$

By choosing FT = 1, we obtain GR = 0.047 FT

Finally,

$$\mathrm{GR} = 0.0235\ \text{ṢR (diameter ŠFH̱)} \qquad (16)$$

Ibn al-Haytham's result differs only slightly from this value ($\frac{1}{40} = 0.0250$). His approximation makes sense. However, the approach just described is not fully satisfactory because it faces a further difficulty.

The demonstration supplied by Ibn al-Haytham aims at providing a criterion to define the threshold at which the projected image appears as a circle—to be precise: it does not differ much from a circle—a fact that is meant to apply when line GR is less than two percent of the diameter ṢR of circle ŠFH̱ (Fig. 4.7).

---

26. Sagitta is another name for versine. As the circle ŠG with center K and radius ŠK is not explicitly stated as a unit circle in the text, I prefer to use the broader term "sagitta."

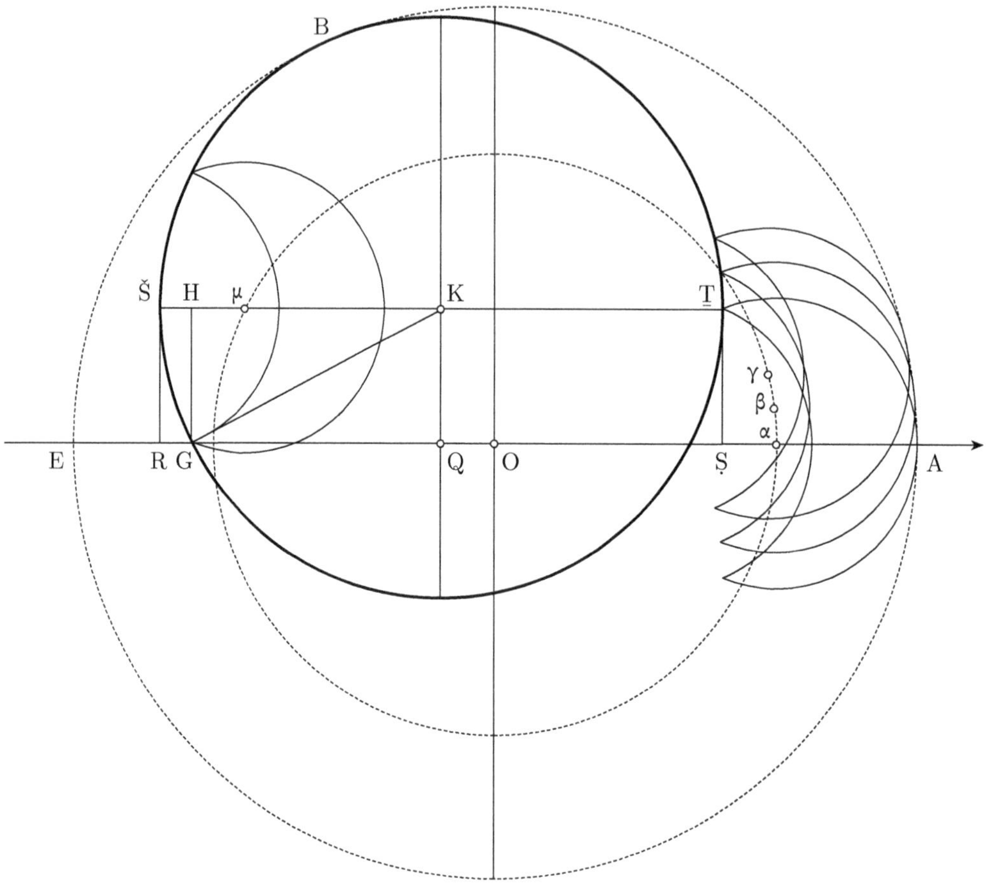

Fig. 4.7. The Determination of Ratio GE/AE

The flattening of the image is expressed as ratio GR/ṢR. However, circle T̲BŠG of diameter ŠT̲ delineating the crescents is not the perimeter of the image. Neither is the case that "the curve of the convex arc [ABE] overlaps with the arc passing through the two points Š G, because this arc is a part of the circle of center K whose perimeter is a part of the convex arc" (lines 636–8). Let us explain this in detail.

When the center of the crescent moves along the positions α, β, γ ... of the dotted circle of radius OṢ centered on O, the perimeter of light protrudes on the right side of circle T̲BŠG of a distance AṢ, which is made of Aα = FT, the radius of the arc of the crescent, and αṢ, the depth of the cavity of any single crescent.

Similarly, when the crescent reaches the opposite side of the circle, the innermost point G of the image is obtained when the ordinate of the center is $\mu = \text{KQ}$. On the other hand, a perfectly circular image would appear if the Sun was not eclipsed, and it would be defined by the successive points of tangency of the solar disc image, when its center moves along the said circle. This perimeter is line ABE.

The difference between the empirical and theoretical images boils down to the difference between the perimeter ABŠG (which is mirror-symmetrical to the axis Ox) and the outer circle ABE. This difference should be expressed as the ratio of GE/AE.

On the one hand, $\text{AE} = 2\,(R + r) = 22$ FT.

On the other hand, $\text{GE} = \text{GR} + \text{RT} + \text{TE}$. $\text{TE} = q = 2r(1-g) = \frac{1}{2}\,\text{FT}$, therefore $\text{TR} = q/2 = \frac{1}{4}\,\text{FT}$ and $\text{GR} = 0.047$ FT (16), hence $\text{GE} = 0.797$ FT.

The searched ratio is $\text{GE}/\text{AE} = 0.797/22 = 0.036$. This value, albeit not much different from that provided by Ibn al-Haytham, involves a comparison of line GE to the diameter of the circle ABE that has little to do with comparing line GR to the diameter of circle ṮBŠ, just as Ibn al-Haytham did in his own work.

4. The Rounding of the Image

In between the two thresholds, the solar image is neither crescent-shaped, nor circular, but simply 'rounded.' This boils down to the following condition

$$R_0\,\ell_1/\ell_0 \leqslant R \leqslant 10\,R_0\,\ell_1/\ell_0$$

5. Stability of these Conditions

As the previous conditions derive from astronomical estimates that have changed over time, we may wonder about the stability of these results, by using either the astronomical values known to Ibn al-Haytham or the modern ones. It then turns out that ratio R_0/ℓ_0—which is similar to an angular diameter—did not vary much from antiquity; thus one may expect that the previous conditions show little sensitivity to the variation of astronomical estimates.

This can be checked by comparing the estimates of the distances and sizes of the celestial bodies provided by Ptolemy (Van Helden 1985: 27) and by al-Farghānī (Van Helden 1985: 30) and modern research (Simon et al., 1998: 24).

Estimate	Ptolemy (Earth radii)	al-Farghānī (Earth radii)	Current Values (AU)
Sun's radius R_0	$5\frac{1}{2}$	$5\frac{1}{2}$	0.004,691
Sun-Earth distance ℓ_0	1,210	1,170	1
Crescent-shaped if $R \leqslant$	14.05 mm	14.53 mm	14.50 mm
Circular if $R \geqslant$	140.5 mm	145.3 mm	145.0 mm

As the thresholds calculated from Ptolemy's, al-Farghānī's and current estimates match well with one another, these conditions are stable and may be used to describe the shape of the image. In Fig. 4.8, each threshold is expressed in function of the R and r magnitudes.

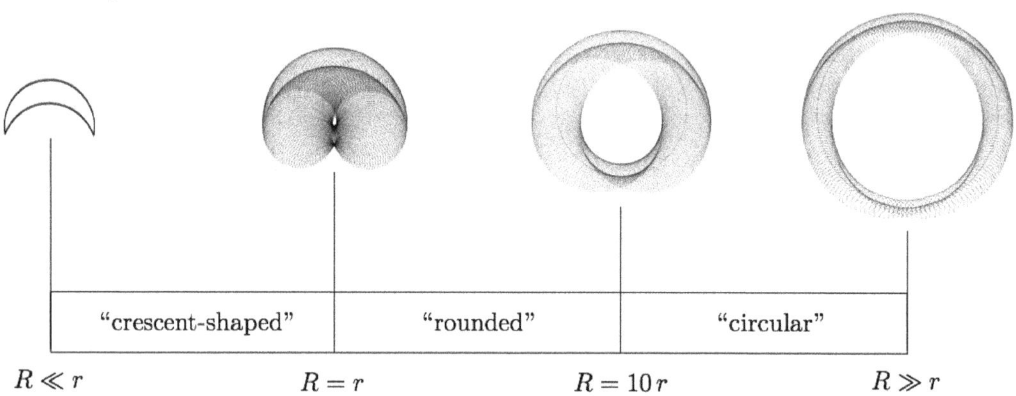

Fig. 4.8. Thresholds for the Deformation of the Image

4. The Image as a Function of the Size of the Aperture

Naẓīf (1942: 192; reed. 2008: 286) provided a first comment on the shape and dimension of the perimeter of the patch of light obtained through wide apertures. Sabra (1972: 196) reproduced his diagram with minor changes to the lettering and basically repeated his analysis (Fig. 4.9).

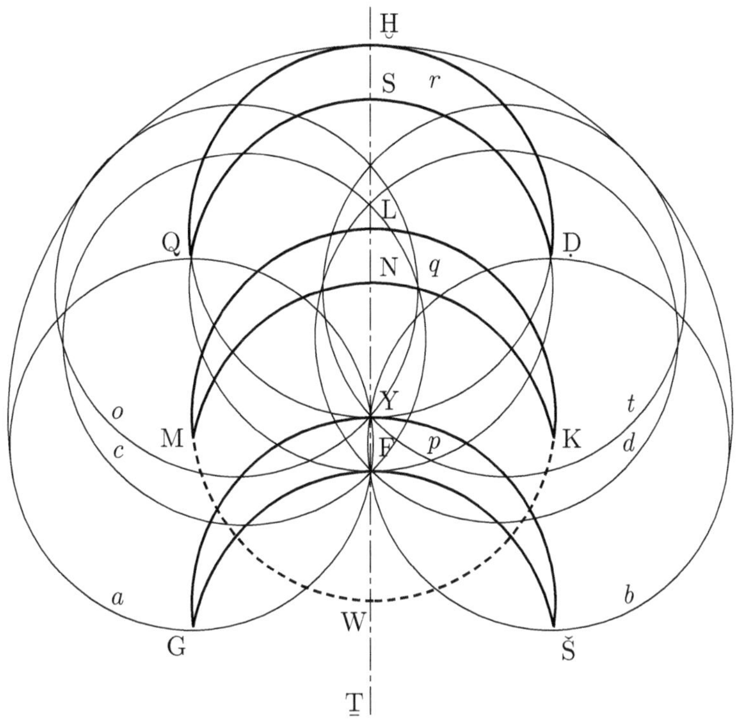

Fig. 4.9. Naẓīf's study of the Perimeter of Light[27]

This diagram shows the perimeter of light H̱ŠFG, illustrating the rounding of the image when the size of the aperture is increased. Sabra writes:

> The crescents *p, q, r* [i.e., the lower, mid and upper central crescents ŠYGF, KLMN and ḌH̱QS, in Naẓīf's lettering] are inverted images produced by three double conical solids of light whose vertices are three different points of the aperture, and whose bases are, on the one side, the shining solar crescent, and, on the other, the inverted image ...
> The center of each circle is therefore the point at which the axis of the cone, passing through the center of the aperture, intersects the screen. It is clear that the centers of all circles will be points on crescent *q* [i.e., KLMN] and that their radii, as well as those of the arcs forming crescents *p, q, r*, will all be equal. The resultant image will therefore be bounded from above by a convex curve of which the upper part is the tangential arc of a circle whose center is the midpoint *k* [i.e., H̱] of the convex arc of crescent *p* [i.e., ŠYGF], and whose radius is twice the radius of that arc. Although circles of light will occur below arc GTH [i.e., GFŠ], they will be relatively few" (Sabra 1972: 196).

27. The lettering of this diagram is not Ibn al-Haytham's; rather it belongs to the tradition of Kamāl al-Dīn/Naẓīf/Sabra.

The rounding of the image of light is because any non-point aperture causes a blurring of the patch of light. In Fig. 4.9, point K is seen as circle ḌŠFY; point M is seen as circle QYFG, and the like.

This approach has three minor shortcomings.

Firstly, there is no mention in Naẓīf's and Sabra's comments that the circle along which the crescent-shaped figure moves while describing the perimeter of light.[28]

Secondly, the geometric diagram is false. The line joining the centers of the two circles t d (around points ḌLYK) as well as that joining the centers of the two circles o c (around points LQMY) make an angle with axis H̱T. There is no reason for that since the solar crescent does not rotate in the sky.

Thirdly, Naẓīf and Sabra consider that the circular aperture and crescent-shaped image have the same diameter. This is an arbitrary choice, for many other situations can occur. This choice $R = r$ separates the two cases that will be discussed in the following pages.

Generalizing from Naẓīf's and Sabra's commentaries, we may study the shape and dimension of the crescent-shaped image by varying the size of the aperture. Consider a dark chamber equipped with a circular aperture of radius R. We assume the Sun is virtually at infinity, so the solar rays are parallel and the light cast on the projection plane has the same radius R as the aperture. Two equivalent methods may be used to investigate the geometry of the image.

Method 1. As stated in *On the Shape of the Eclipse*: "If the apex of the cone moves along the perimeter of the aperture ... the endpoint of the axis, which is in the plane parallel to the aperture, will describe a perimeter of a circle parallel to the perimeter of the aperture" (line 374–9). The light of the whole solar disc AǦ focuses in one point of the aperture—the apex H—and is cast on the projection plane as circle KL of radius r. When the apex H of the cone moves along the perimeter of aperture HḤ, circle KL describes the whole circumference KM of the image (Fig. 4.10).

28. The circle of center Y, which is the projection of the circular aperture is not drawn in Naẓīf's and Sabra's diagrams. This is the dotted line LKWM added to Fig. 4.9.

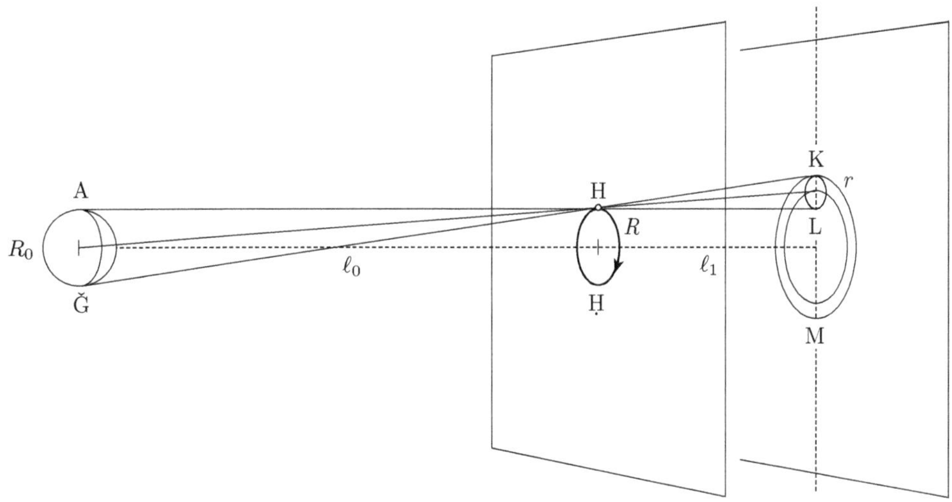

Fig. 4.10. Projection Method 1

Method 2. Equivalently, the light cone originates in one point of the solar disc AǦ and radiates through the whole aperture HḤ toward circle KL of radius R. When point A moves along the perimeter of the Sun AǦ, circle KL describes the whole circumference KM of the image (Fig. 4.11).

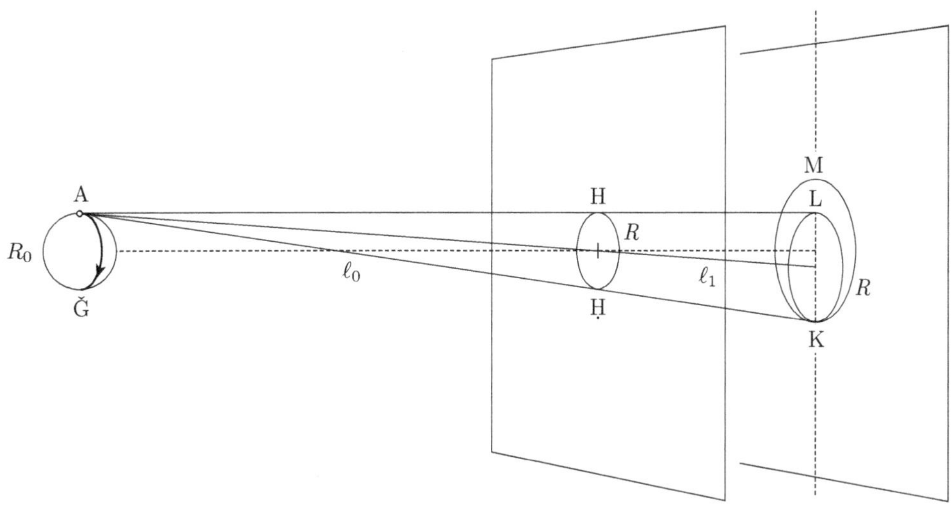

Fig. 4.11. Projection Method 2

Now let some crescent-shaped image be represented on the orthonormal basis Oxy with method 1 (Plate 4.1, p. 250). For clarity, the lines are shown on Fig. 4.12.

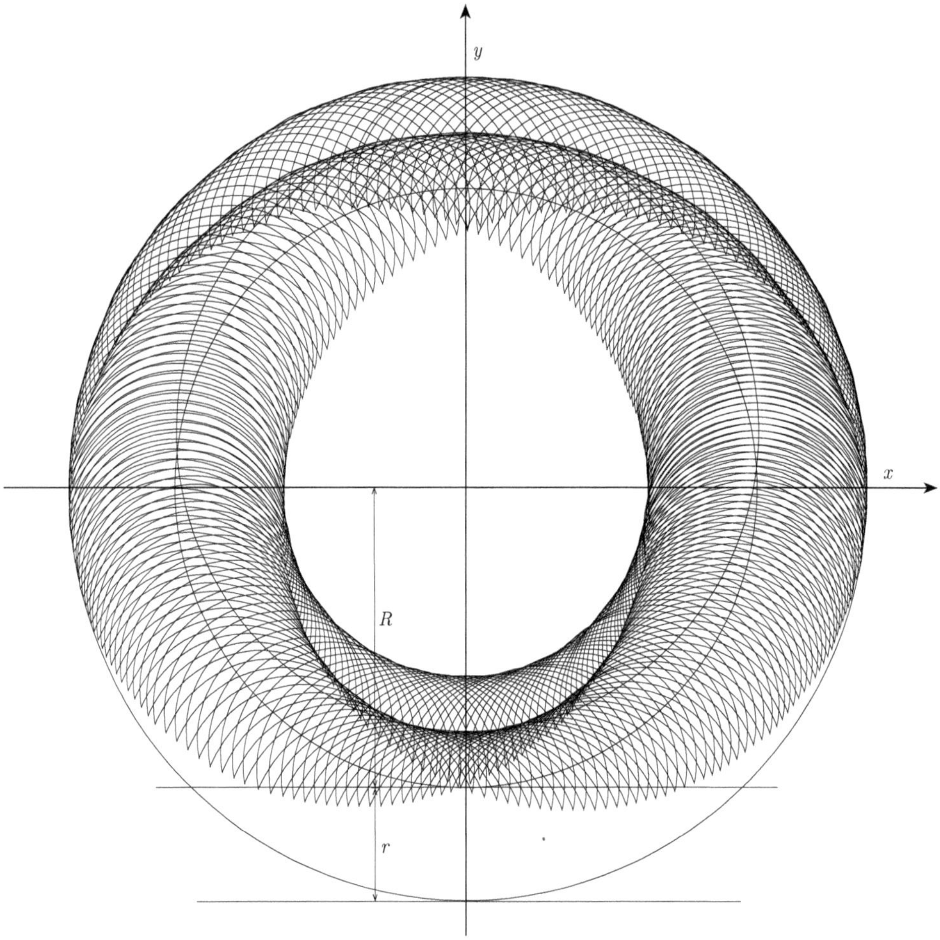

Fig. 4.12. Point Analysis of a Crescent-Shaped Image (Geometric Lines)

Ibn al-Haytham's reasoning is better explained in modern terms. The geometry of the image depends on several factors (Fig. 4.13): q denotes the distance from the solar limb to the lunar limb; p denotes the distance from the lunar limb to the center of the Sun; n is the height of the crescent tips from the center of the Sun.

Numbers q and n can be written in function of the eclipse magnitude. If the Sun and the Moon have the same apparent diameter $2r$, the magnitude of an eclipse (that is, the fraction of the solar diameter eclipsed by the Moon), equals $g = (2r - q)/2r$. Thus $q = 2r(1 - g)$. However, $n = q/2$; therefore $n = r(1 - g)$. Let us now discuss the two cases $R < r$ and $R \geqslant r$.

Case 1

When $R < r$ the perimeter of light is the curve ABCD, which consists of the three arcs AB, BC and CD (Fig. 4.13).

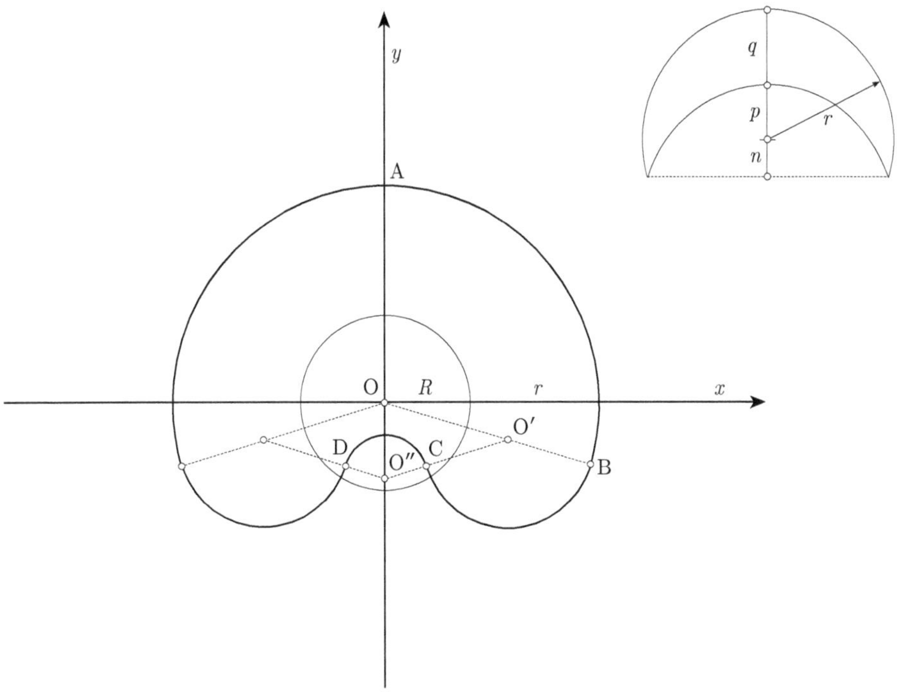

Fig. 4.13. The Perimeter of Light when $R < r$

Arc AB is such that:

$$\text{Center O}\begin{cases} x = 0 \\ y = 0 \end{cases}$$

Radius $R_{AB} = R + r$

As O′ is on the circle of center O and radius r, $x = \sqrt{r^2 - n^2} = \sqrt{r^2 - r^2(1-g)^2}$, which reduces to $x = r\sqrt{g(2-g)}$. Thus arc BC is such that:

$$\text{Center O}'\begin{cases} x = \pm r\sqrt{g(2-g)} \\ y = -r(1-g) \end{cases}$$

Radius $R_{BC} = R$

Finally, arc CD is such that:

$$\text{Center O}'' \begin{cases} x = 0 \\ y = -2r(1-g) \end{cases}$$

Radius $R_{CD} = \text{O}''\text{O}' - \text{CO}'$. $\text{O}''\text{O}' = \text{OO}' = r$ and $\text{CO}' = \text{O}'\text{B} = R$, thus:

Radius $R_{CD} = r - R$

Arcs AB and BC are tangent at point B, which is aligned with points O and O′. Arcs BC and CD are tangent at point C, aligned with O′ and O″.

Case 2

When $R \geqslant r$ the perimeter of light is the curve ABC—not to be confused with a cardioid. This curve consists of two arcs (Fig. 4.14).

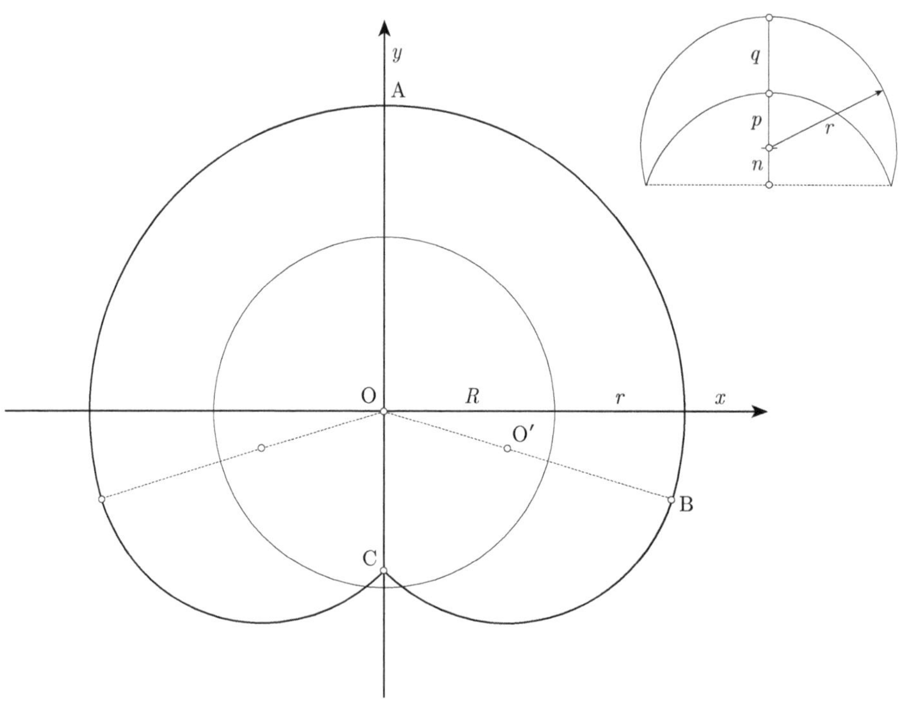

Fig. 4.14. The Perimeter of Light when $R \geqslant r$

Arc AB is such that:

$$\text{Center O}\begin{cases} x = 0 \\ y = 0 \end{cases}$$

Radius $R_{AB} = R + r$

Arc BC is such that:

$$\text{Center O}'\begin{cases} x = \pm r\sqrt{g(2-g)} \\ y = -r(1-g) \end{cases}$$

Radius $R_{BC} = R$

The two arcs are tangent at point B, which is aligned with points O and O′. The coordinates of cusp C, which is the intersection of the curve with axis Oy, can be determined in the following way (Fig. 4.15).

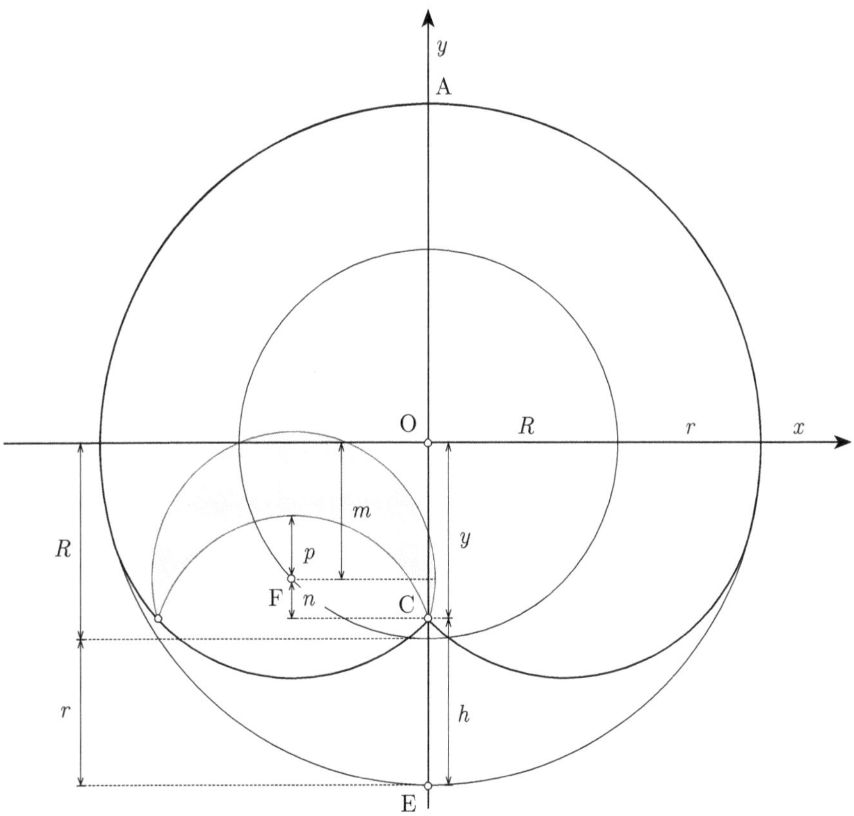

Fig. 4.15. The Coordinates of the Cusp

Cusp C belongs to both the axis Oy and the circle of radius R, whose center is point F with abscissa $-r\sqrt{g(2-g)}$ and ordinate $-m$. The equation of the circle fixes $m=\sqrt{R^2-r^2g(2-g)}$. The ordinate of the cusp is $y=-(n+\sqrt{R^2-r^2g(2-g)})$. However, $n=r(1-g)$. Therefore point C has the coordinates:

$$\text{Cusp C}\begin{cases} x=0 \\ y=-r(1-g)-\sqrt{R^2-r^2g(2-g)} \end{cases}$$

3. Flattening

As the deformation of the image mainly consists of the cavity CE (Fig. 4.15), the ratio CE/AE is the simplest way to express the flattening of the image (Chapter 4.3, pp. 127–8). The ratio is $F=h/2(R+r)$; $0 \leqslant F \leqslant 1$, with $h=\text{CE}=(R+r-y)$. By replacing the y value in h, we obtain $h=R+rg-\sqrt{R^2-r^2g(2-g)}$.

Therefore

$$F=\frac{R+rg-\sqrt{R^2-r^2g(2-g)}}{2(R+r)}$$

This equation can be simplified by introducing the ratio $k=R/r$. Function F will then be determined by only two variables, g (the magnitude of the eclipse) and k (the ratio of the radius of the aperture to the radius of the crescent).

Let us introduce k,

$$F=\frac{rk+rg-\sqrt{r^2k^2-r^2g(2-g)}}{2(rk+r)}$$

Then factorize,

$$F=\frac{r(k+g)-r\sqrt{k^2-g(2-g)}}{2r(k+1)}$$

And eliminate r,

$$F=\frac{k+g-\sqrt{k^2-g(2-g)}}{2(k+1)}$$

In the graph below, we plot the flattening of the image F in ordinates versus ratio k in abscissae, for different magnitudes of the eclipse ($0 \leqslant g \leqslant 1$) (Fig. 4.16).

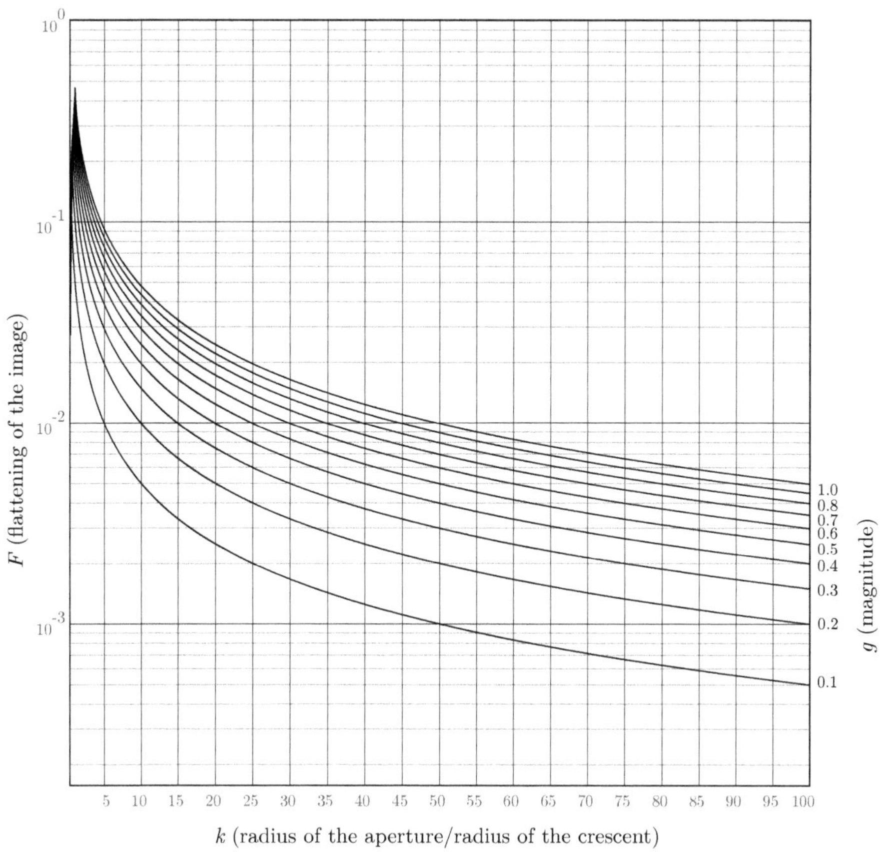

Fig. 4.16. The Flattening of the Image as a Function of g and k

Example. Consider Ibn al-Haytham's value for the threshold of a 'circular' image, $g = 3/4$ and $R = 10\,r$ (lines 536–41).

The flattening of the image is

$$F = \frac{k + g - \sqrt{k^2 - g(2-g)}}{2(k+1)} = \frac{10 + \frac{3}{4} - \sqrt{100 - \frac{3}{4}\left(\frac{5}{4}\right)}}{2(10+1)}$$

that is

$$\frac{\frac{43}{4} - \sqrt{\frac{1585}{16}}}{22} \approx 0.036$$

Equivalently, this number may be read directly in Fig. 4.16. Consequently, when the radius of the aperture is ten times the radius of the crescent, the flattening of the image is about three percent of the diameter of the circular image. This is the tolerance permitted by the text.

4. Special Cases

There are three cases worth considering:

1) When $R = 0$, the aperture reduces to a point. The perimeter of light is a perfect crescent-shaped image. We reach the same result as in Chapter 3.6, p. 104.

2) When $R = r$, the result is that of Naẓīf and Sabra. The perimeter of light is the curve ABC, which consists of two arcs (Fig. 4.17).

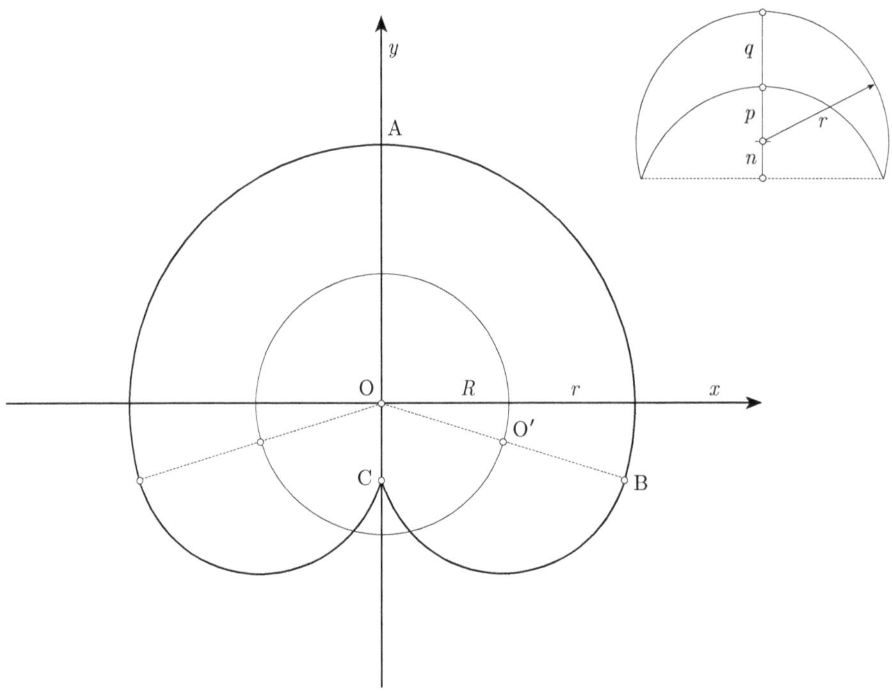

Fig. 4.17. The Perimeter of Light when $R = r$

Arc AB is such that:

$$\text{Center O}\begin{cases} x = 0 \\ y = 0 \end{cases}$$

Radius $R_{AB} = 2R = 2r$

Arc BC is such that:

Center O' $\begin{cases} x = \pm r\sqrt{g(2-g)} \\ y = -r(1-g) \end{cases}$

Radius $R_{BC} = R = r$

Arc CD seen above (Fig. 4.13) is such that:

Radius $R_{CD} = r - R = 0$

Thus circle CD reduces to a point; this is the cusp C = O″ of coordinates:

Cusp C $\begin{cases} x = 0 \\ y = -2r(1-g) \end{cases}$

3) Then, as the aperture is widening, the perimeter of light is gradually rounding. However this curve will never become a perfect circle because this would imply that $R = \infty$, a condition which is never fulfilled.

With a few exceptions, Ibn al-Haytham's text is clear on the point. It is only in Section 1.1 (*The Observations*) that the image is said to be circular:

> "The more increased the capacity of the aperture, the more changed the image. This change continues until some limit of capacity of the aperture, at which the figure that appears in the light emerging from the aperture is no longer crescent-shaped, but circular (الضوء الذي يخرج من الثقب مستديرًا)" (lines 23–5).

However, in the main text, Ibn al-Haytham provides a more nuanced view on the exact shape of the projected image. In Section 4.1 (*Effect of the Size of the Aperture on the Concave Face*), he states:

> The form of light that appears on the plane parallel to the aperture is a crescent-shaped figure, whose convex arc is of perfect circularity, and concave arc is not of perfect circularity, though it looks round to the sense (lines 488–9).

And, a few lines further down:

> The form [of the sunlight] is crescent-shaped, its convexity is round of perfect circularity, while its concavity is round to the sense only (lines 521–2).

Even though the condition $R = \infty$ is not specified, we come to the conclusion that the text expressly acknowledges that the curve delineating the image is not a perfect circle on the concave side of the image.

5. The Image as a Function of the Focal Distance

In Section 4.2 (*Effect of the Distance on the Image of the Concave Face*), Ibn al-Haytham studies the shape and dimension of the perimeter of light as a function of the focal distance.

Let R be the radius of the aperture and let r be the radius of the crescent. Since the Sun is virtually at infinity, the rays passing through the aperture extend parallel to each other. The patch of light will have the radius R of the aperture, whatever the focal distance of the dark chamber. On the contrary, the radius r of the crescent will vary with the focal distance of the darkroom, because its size is fixed in two opposite similar triangles, whose common vertex is a point of the aperture, and whose bases are, on the one side, the solar crescent, and, on the other side, the image cast on the projection plane.

The greater the focal distance, the wider the crescent.

1. General Case

The radius r of the crescent is given by the equation $r = R_0\,\ell_1/\ell_0$, where R_0 is the Sun radius, ℓ_0 is the Sun-Earth distance and ℓ_1 is the focal length of the chamber. Let us draw the image on the orthonormal basis Oxy (Fig. 4.18).

When focal length ℓ_1 increases, $r' > r$ according to relation $R_0/\ell_0 = r/\ell_1 = r'/\ell'_1$. As R is constant, the perimeter of light A′B′C′D′ enlarges to A″B″C″D″. (For details on these curves, see Chapter 4.4, pp. 130–42.)

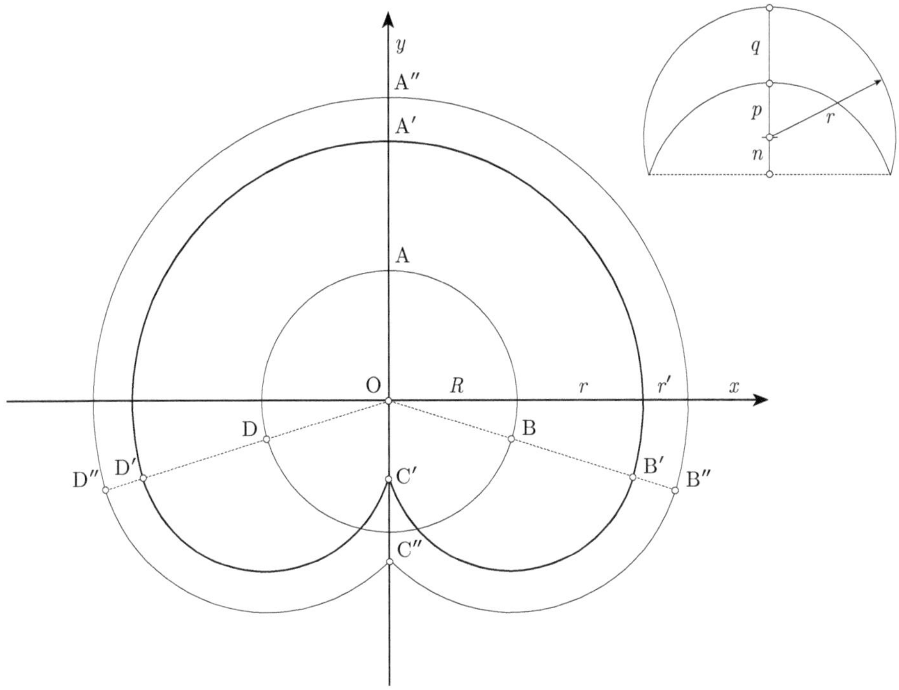

Fig. 4.18. The Perimeter of Light when $r' > r$

2. Flattening

The flattening of the image in function of the focal distance can be studied in the following way. We start with the equation of the flattening (Chapter 4.4, p. 138):

$$F = \frac{R + rg - \sqrt{R^2 - r^2 g(2-g)}}{2(R+r)}$$

We replace $r = R_0\,\ell_1/\ell_0$,

$$F = \frac{R + g\frac{\ell_1}{\ell_0}R_0 - \sqrt{R^2 - g(2-g)\left(\frac{\ell_1}{\ell_0}R_0\right)^2}}{2\left(R + \frac{\ell_1}{\ell_0}R_0\right)}$$

Given R_0, ℓ_0 and $R = 2$ mm, F is a function of variables g (the magnitude of the eclipse) and ℓ_1 (the focal distance, in mm). We plot F in ordinates and focal distance ℓ_1 in abscissae, for different magnitudes of the eclipse ($0 \leqslant g \leqslant 1$) (Fig. 4.19).

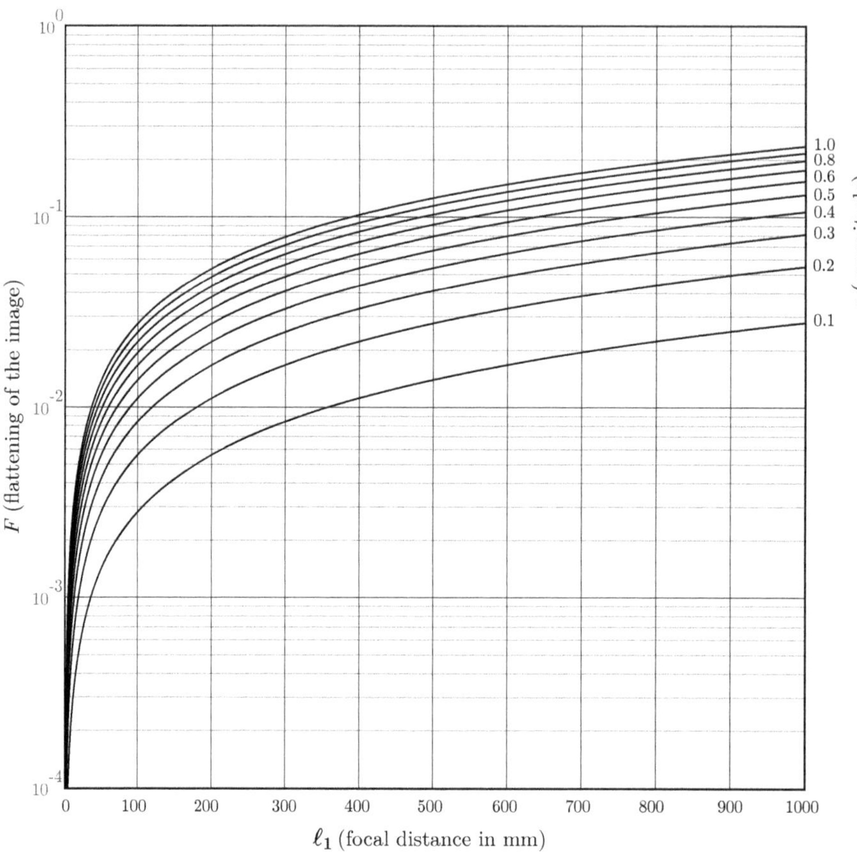

Fig. 4.19. The Flattening of the Image as a Function of g and ℓ_1

3. Special Cases

When the aperture is in contact with the projection plane, $\ell_1 = 0$ and $r = 0$. The patch of light reduces to the inner circle ABD which has the size of the aperture. When the aperture is removed from the projection plane, the patch of light widens in proportion to the distance. The farther the projection plane, the greater the image.

Two different phenomena overlap here: one geometrical, the other physical. The crescent-shaped image is made of the same amount of light, whatever its size. Therefore the relationship set out in the previous paragraph leads to the following one: The farther the plane, the dimmer the light.

This is exactly what Ibn al-Haytham states in his treatise:

"If the aperture is moved far from the projection plane ... The concavity is more sharp than the first concavity, because the intersection of the light circles is under point F, and the luminous parts that are within arc ŠFH̱ are fewer than the lights that are inside the arc of the first concavity. Therefore ... the concavity <subjected> to the light will be increasingly deep" (lines 490–508).

Conversely:

"The reverse occurs if the aperture is moved close to the plane parallel to it ... Therefore the concavity is less <sharp> and smaller <in size>. As a result, whenever the aperture comes closer to the plane that displays its light, the concavity that appears in the light will be less <sharp>" (lines 509–18).

6. The Image as a Function of the Shape of the Aperture

In a third experiment, Ibn al-Haytham wonders whether a modification in the shape of the aperture (circle, square, or whatever) changes the solar image.

1. Graphic Simulation and Experiments

As the geometry of the image has been defined in the previous sections, we can run a graphic simulation to better understand Ibn al-Haytham's results. The image has been calculated for different shapes of the aperture, while varying its size. We have tested a circular light source with a square aperture; a circular light source with a triangular aperture; and a crescent-shaped light source with a circular aperture (Plates 4.2–4, pp. 251–3, where an image is identified by the ratio $k = R/r$). These results have been checked with a modern instrument similar to Scheiner's helioscope (Plate 4.5, p. 254). Some of the images produced appear in Plate 4.6, p. 255. Whatever the method and the shape considered, there is a gradual rounding of the image as the size of the aperture increases. This phenomenon is because radius R of the aperture increases while radius r of the image remains constant.

This conclusion is drawn in *On the Shape of the Eclipse*:

> "If one experiments with sunbeams emerging from wide apertures at the time of the solar eclipse, they will always appear similar to the shape of the apertures: If the wide aperture is circular, the light ... will appear circular; If the aperture is square, the light ... will appear square, and so on, with any other shape of the aperture, provided that it is wide" (lines 29–32).

Note that the major Latin perspectivists, like Bacon (*De speculis comburentibus*), Pecham (*Perspectiva communis*) or Witelo (*Perspectiva*), failed to rediscover these results (Lindberg 1968, 1970*ab*, 1996; Mancha 1989; Thro 1994).

2. Transformation of the Image

We can investigate further the gradual transformation of the image in function of the widening of the aperture. Let us take the case of a circular light source sending a lightbeam through a triangular aperture (Fig. 4.20).

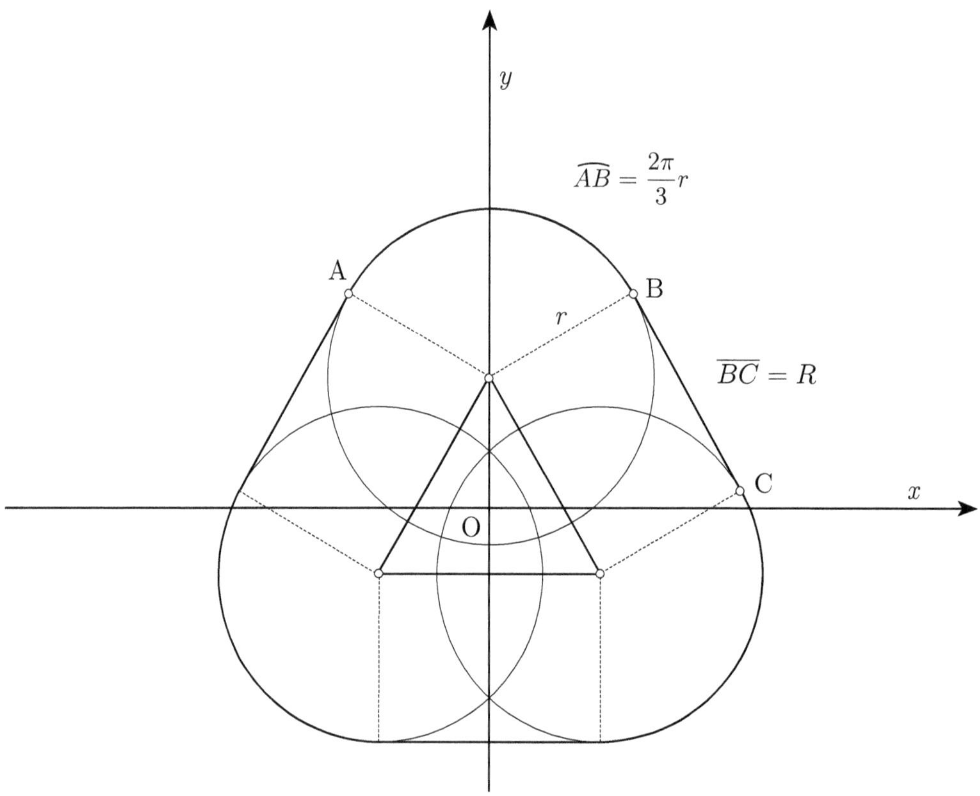

Fig. 4.20. The Image by a Triangular Aperture

When the aperture is a point, the image is a circle; when it has a radius $R \gg r$, it is a triangle. Between the two ends, the image is a triangle of side $\overline{BC} = R$, with angles rounded by an arc $\widehat{AB} = \frac{2\pi}{3} r$. This shape may be expressed by the ratio of the linear perimeter to the whole perimeter. In the case of a n-sided polygon, the linear perimeter is nR, the whole perimeter is $2\pi r + nR$. Thus ratio P_L/P is

$$\frac{P_L}{P} = \frac{nR}{2\pi r + nR}$$

We replace $r = R_0\,\ell_1/\ell_0$ and get

$$\frac{P_L}{P} = \frac{nR}{2\pi\left(\frac{\ell_1 R_0}{\ell_0}\right) + nR}$$

The variation of this ratio describes the change in the image. We plot R in abscissae, P_L/P in ordinates for $r = 14.5$ mm and different n-sided polygons (Fig. 4.21).

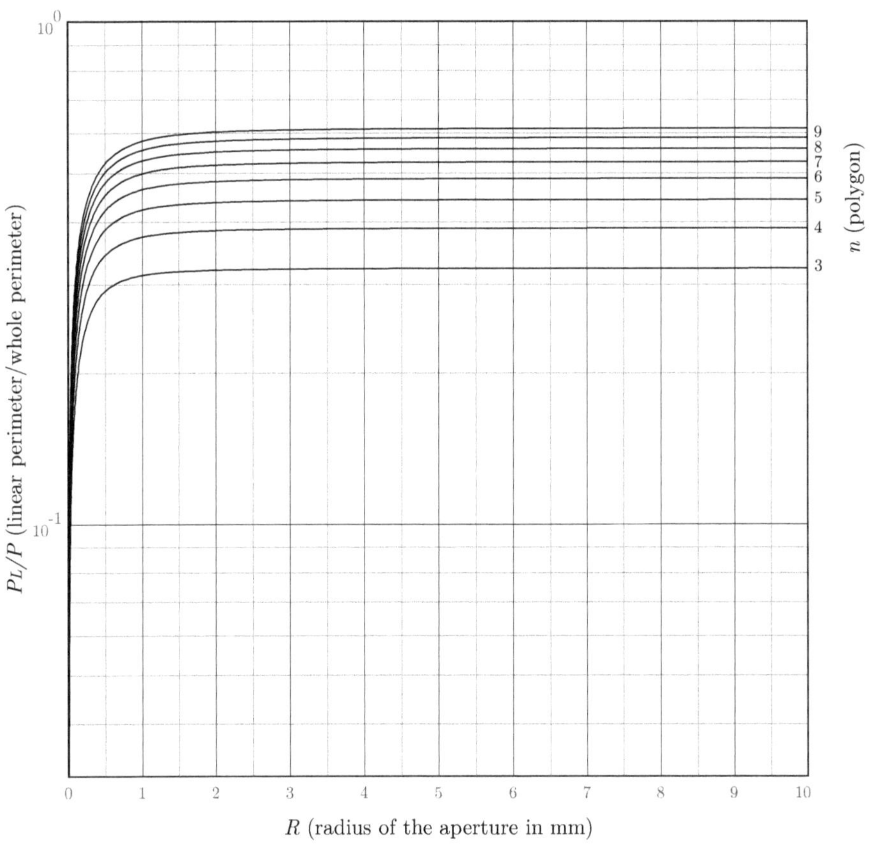

Fig. 4.21. The Ratio P_L/P as Function of R and n

7. The Image as a Function of the Light Source

Despite its title, *On the Shape of the Eclipse* is more an optical work than an astronomical treatise. The choice of studying a solar partial eclipse arises here essentially from optical considerations: the Sun provides a powerful light source; the partial eclipse breaks symmetry of the solar disc.

In the final Section 5 (*Analysis of the Image in the Case of the Moon*), Ibn al-Haytham pushes his experimentation one step further by comparing the images of the solar and lunar crescents. This allows him to change the distance and intensity of the light source. He demonstrates by two parallel lines of reasoning that the Moon never appears crescent-shaped.

1. Geometry

Ibn al-Haytham declares that the condition $R/R_0 \leqslant \ell_1/\ell_0$ for an image to be crescent-shaped is not met by the Moon for geometrical reasons. He first refers to Ptolemy's datum: "the diameter of the Sun is $18\frac{4}{5}$ the diameter of the Moon" (lines 656–7; *Almagest* V, 16; Halma 1813: 347; Toomer 1984: 257). The Sun and the Moon have the same angular diameter, so $18\frac{4}{5}$ is also the ratio of distances. By similar triangles, the diameter of the aperture and the focal length of the chamber must be $18\frac{4}{5}$ smaller than the corresponding dimensions in the case of the Sun (lines 661–8). In these circumstances, Ibn al-Haytham claims, $R/R_0 \leqslant \ell_1/\ell_0$ cannot be met from the Earth: "[As regard] the ratio required for the moonlight to appear crescent-shaped ... perhaps one cannot find on the face of the Earth any place whose distance to the aperture meets the required ratio" (lines 702–4).

The core of the argument is found when Ibn al-Haytham declares:

> "If the Moon is facing an aperture like that by which the sunlight appears crescent-shaped, and if the moonlight passes through it and attains the plane whose distance from the aperture is one part out of eighteen times the distance that shows the sunlight crescent-shaped, then the light will appear circular" (lines 685–8).

This statement may be checked by using data provided by Ptolemy (Halma 1813; Toomer 1984) and modern research (Simon et al., 1998).

Estimate	Ptolemy (Earth radii)	Ptolemy (Sun radii)	Current Values (km)
Sun's radius $R_{\odot}$	$5\frac{1}{2}$	1	695,990
Sun-Earth distance $\ell_{\odot}$	1,210	220	149,597,870
Moon's radius $R_{☽}$	$\frac{1}{4}+\frac{1}{24}$	$1/18\frac{4}{5}$	1,737.4
Moon-Earth distance $\ell_{☽}$	59	$10\frac{3}{4}$	383,398

In the case of the Sun, the image is crescent-shaped if $R \leqslant R_{\odot}\ell_1/\ell_{\odot}$, which leads to $R \leqslant 14.05$ mm (Ptolemy).

In the case of the Moon, we retain this value of R, we use $R_{☽}$ and $\ell_{☽}$ instead, and give ℓ_1 $1/18\frac{4}{5}$ the value of the focal distance when the Sun is observed. We now may wonder whether $R \leqslant \ell_1 R_{☽}/\ell_{☽}$.

$$\frac{\ell_1 R_{☽}}{\ell_{☽}} = \frac{\frac{3091}{18\frac{4}{5}}\left(\frac{1}{4}+\frac{1}{24}\right)}{59} = 0.81\,\text{mm}$$

Therefore,

$$R \geqslant \ell_1 R_{☽}/\ell_{☽}$$

In addition, Ibn al-Haytham notes that the condition for an image to be circular is satisfied in this case, because

$$R \geqslant 10\, R_{☽}\, \ell_1/\ell_{☽}$$
$$(14.05\,\text{mm} \geqslant 10 \cdot 0.81\,\text{mm})$$

Ibn al-Haytham's numerical statement[29] is correct, but is based on a flaw. Indeed there is no reason why, after considering a homothety (lines 663–8), Ibn al-Haytham applies the $18\frac{4}{5}$ coefficient only to the focal length of the chamber and not to the size of the aperture. When homothety is applied consistently, the condition for an image to be crescent-scaped is, in fact, fulfiled, because $18\frac{4}{5}$ is applied to the expressions situated on both sides of the inequals sign.

29. There is no need to proceed with current estimates, because the ratio $R_{☽}/R_{\odot} = 1/400$ gives the aperture a size much smaller.

By only considering geometrical aspects, the conclusion is reached that the Moon does cast a crescent-shaped image, just as does the Sun. The geometrical argument is thus flawed.

2. Proto-Photometry

What saves Ibn al-Haytham's geometrical reasoning is that it is mixed with physical considerations which reflect photometry *avant la lettre*. Point analysis of light offers a simple way to estimate the quantity of light which is included in the image: this is the sum of the patches of light overlapped in a given region.

As a first step, it is noted that condition for the image to be crescent-shaped requires an aperture smaller than that used with the Sun. But a diameter $18\frac{4}{5}$ smaller than one "grain of barley," which is about 0.1 mm wide, virtually reduces to a point. When the diameter of the aperture is one part out of $18\frac{4}{5}$ of that facing the Sun, its area is $\frac{1}{324}$ of that facing the Sun. Such a small aperture transmits so little light that it can be considered to be "imperceptible, especially the light of the Moon, which is faint" (lines 675–6). A second factor prevents the observation of an image: the Moon has little light. The Moon's illuminance (10^{-1} lux)—which denotes the total luminous flux of the source per unit area—is one million times lower than that of the Sun (10^5 lux).

However, Ibn al-Haytham does not stop here. He then reasons as if the Moon was luminous enough to cast a crescent-shaped image on the projection plane:

> "Assuming that the light coming from the light of the Moon was crescent-shaped, its concavity would disappear with the remoteness of the aperture, because the parts of the light that weaken first with the remoteness are the angles and edges. If what had made the moonlight appear crescent-shaped at a short distance moved away from the aperture, then it is its angles which would vanish first, then its edges, so that it would become circular" (lines 692–5).

Ibn al-Haytham's thesis that the tips and the edges vanish first may be tested by using modern methods, but let us return to the text beforehand.

In Section 2.4 (*The Archimedean Analysis*) the density of the solar image, either circular or crescent-shaped, is explained by the aggregation of an infinity of lights overlapping one another on the projection plane. The patches of light have the following characteristic features:

> "Therefore, on the plane parallel to the aperture, there will be circular lights, compact (متراصّ) and overlapping (متداخل) each other, none of which is distinguished from the rest" (lines 74–5).
>
> "Light, which appears on the plane parallel to the aperture at the time of a solar eclipse, is an aggregate (مركب) of light crescents, contiguous (متصل) and overlapping (متداخل) each other, none of which is distinguished from the rest" (lines 108–10).
>
> "The light circles, whose centers belong to arc KLM, are compact (متراصّ) and interlocked (متصل) with each other" (lines 331–2).
>
> "These circles are interlocked (متصل), compact (متراصّ) and overlap (متداخل) one another from point Š to point Ḫ" (lines 455–6).

On the Shape of the Eclipse resembles *On the Qualities of Shadows* in one respect at least: both texts introduce infinitesimal quantities. While the variation of light and shade within the penumbra is explained by cutting the light source in an infinity of segments, the variation of light within the image is explained by the overlapping of an infinity of patches of lights. Both texts are distant echoes of Archimedes' infinitesimal analysis revived in Medieval Islam through the works of the Banū Mūsā, Thābit Ibn Qurra and Ibn Sinān.

The fact is first evidenced by terminology that evokes continuity. The overlapping lights are called *muttaṣil* (root: *waṣala*), a term that means "contiguous, uninterrupted, continuous." The lights are also called *mutarāṣṣ* (root: *raṣṣa*), which means "compact, dense, tight". And these lights are not separated from each other as Ibn al-Haytham himself declares: "none of which is distinguished from the rest."

Secondly, the fact is confirmed by the mechanical means involved in the proof. As the movement of the apex of the cone along the perimeter of the aperture is *continuous*, it creates an *infinity* of lights on the projection plane.

This recommends using integral calculus to determine the amount of light in each point of the image.

Let us first draw a circular image on a square grid of unit $x_1 = y_1 = 1$ on the projection plane Oxy (Plate 4.7, p. 255). The distribution of light in this image can be studied by looking at the lines underlying this diagram (Fig. 4.22). It is apparent that any point P of the image is illuminated by a circle of radius r, provided its center is on the frame at a distance $d \leqslant r$ from P.

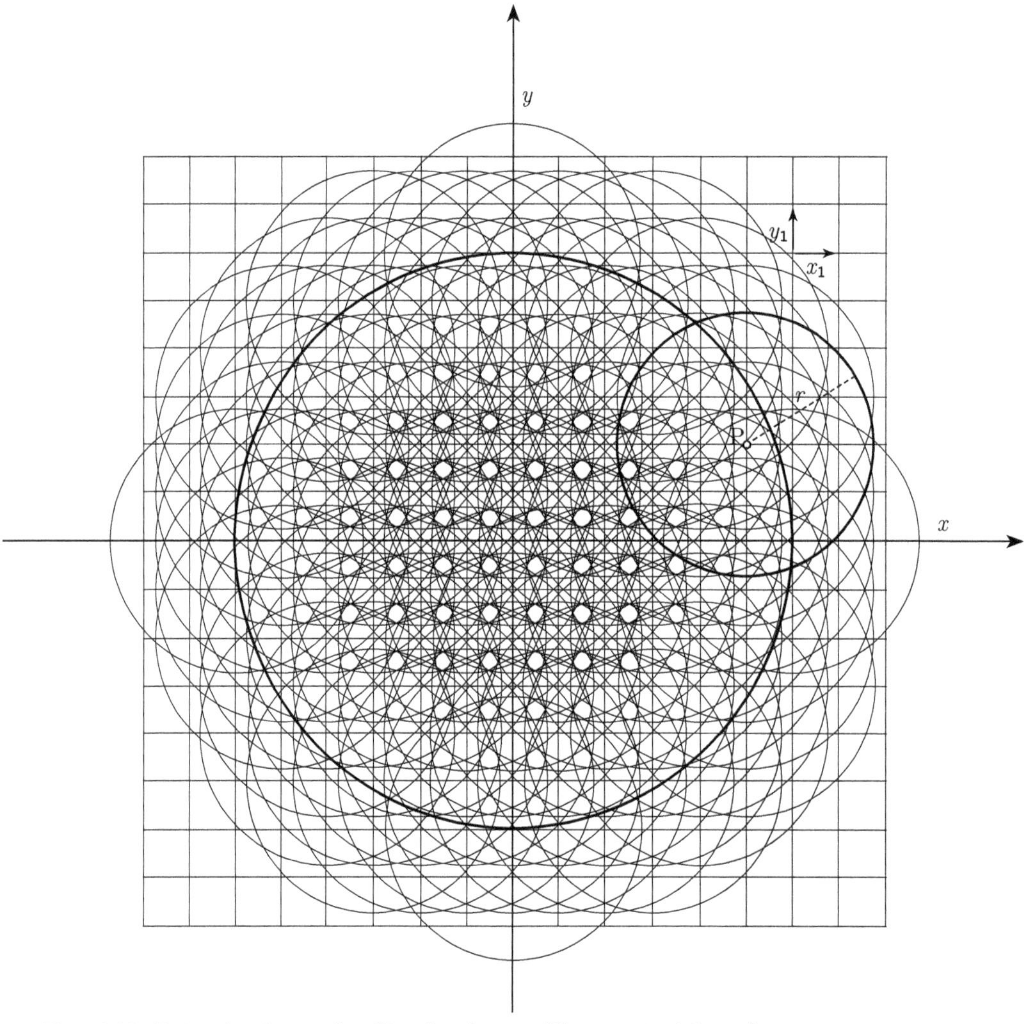

Fig. 4.22. Point Analysis of a Circular Image (Geometrical Lines)

Then integral calculus is used to determine the total amount of light in point P, when P scans the image. Consider a partial solar eclipse of magnitude $m = \frac{2}{3}$, with $r = 4R$, and let the frame unit tend towards zero, $x_1 = y_1 \longrightarrow 0$.

There are three situations to be studied: the center, edge and tip of the crescent (Fig. 4.23).

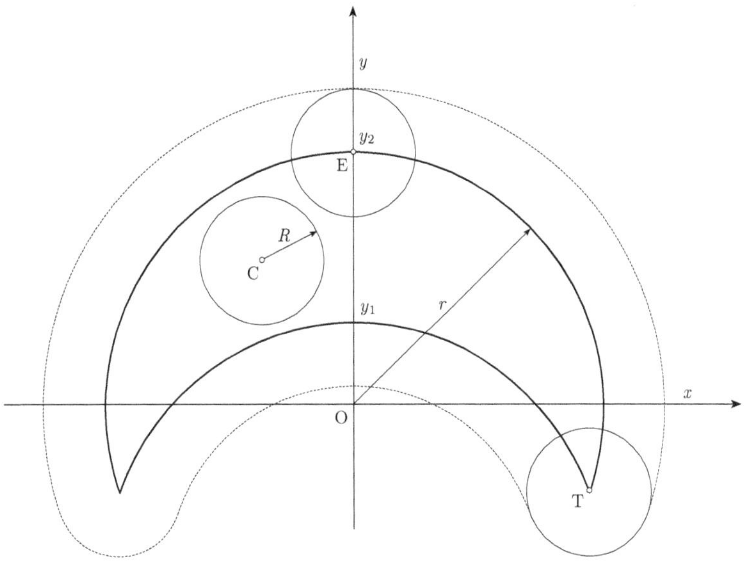

Fig. 4.23. The Center, Edge and Tip of the Crescent

Any point P which is in C (C for center) receives light from all surrounding circular lights whose centers are situated at a distance $d \leqslant R$. When point P moves from the central region of the crescent outward, light decreases and drops to zero when it coincides with any point of the outer dotted line. Therefore, there is no sense to compare the intensity of light of two points *exactly* situated on the edge and the tip, because the result is simply $I_E^* = I_T^* = 0$. This recommends the comparison of the amount of light received by two points situated at some distance from the outer perimeter, such as point E (E for edge) and point T (T for tip), both situated on the inner crescent. Ibn al-Haytham's conclusion that the image vanishes first in T, then in E, and then in P, may be clarified by calculation.

1. Center. Point C, which is located in the central region of the crescent, receives light from all surrounding circular patches of light whose centers are at a distance $d \leqslant R$. Light received in C is the sum of these overlapping patches of light. This sum is the number of circles having their centers within a circle of radius R, i.e., the area of the disc of radius R.

In polar coordinates (θ, r), this is written as

$$\mathscr{A} = \iint_D f(\theta, r)\, d\theta\, dr = \int_0^{2\pi} d\theta \int_0^R r\, dr = \pi R^2$$

The intensity of light is constant on Oy, from the point of ordinate $y_1 + R$ to the point of ordinate $y_2 - R$. The intensity of light is maximum there: $I_C = \mathscr{A} = \pi R^2$.

2. Edge. When the scanning point arrives at point E on the edge of the crescent, it is lit by all surrounding circular lights whose centers are situated in the lens-shaped region ABE. As above, the number of circular lights which lit point E equals the area of this region (Fig. 4.24).

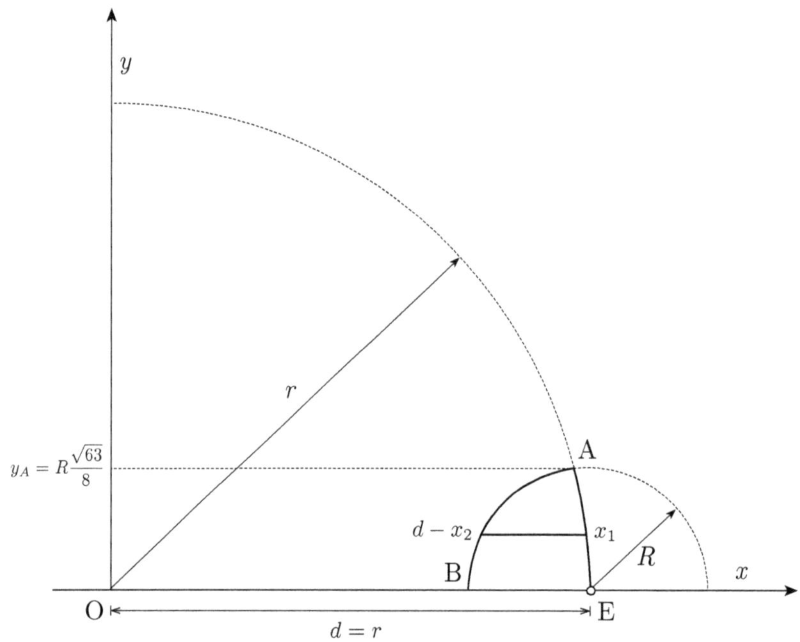

Fig. 4.24. The Amount of Light at the Edge of the Crescent

The approach involved in determining the area of the lens is basically to integrate the difference of functions x_1 and $d - x_2$ by varying y from $y = 0$ to y_A.

The equation of the circle of center O is $x^2 + y^2 = r^2$ and that of the small circle of center E is $(d - x)^2 + y^2 = R^2$. The two circles intersect in two points, A and A′. We select that of positive ordinate $y_A = R\frac{\sqrt{63}}{8}$.

The area of the lens is twice the area of ABE:

$$\mathscr{A} = 2\int_0^{R\frac{\sqrt{63}}{8}} \left[\sqrt{r^2-y^2} - d + \sqrt{R^2-y^2}\right] dy$$

$$= \underbrace{2\int_0^{R\frac{\sqrt{63}}{8}} \sqrt{r^2-y^2}\, dy}_{(1)} - \underbrace{2\int_0^{R\frac{\sqrt{63}}{8}} r\, dy}_{(2)} + \underbrace{2\int_0^{R\frac{\sqrt{63}}{8}} \sqrt{R^2-y^2}\, dy}_{(3)}$$

- The integral (2) is simply

$$2\int_0^{R\frac{\sqrt{63}}{8}} r\, dy = 2\, ry + C$$

Since $F(0) = 0$, we obtain $2\left[ry\right]_0^{R\frac{\sqrt{63}}{8}} = 2\, rR\frac{\sqrt{63}}{8} = \sqrt{63}R^2$

Therefore

$$(2) \;=\; \sqrt{63}R^2$$

- Calculating integral (1) requires the substitution $y = r\,\sin\theta$ and $dy = r\,\cos\theta\, d\theta$. This allows us to rewrite $\sqrt{r^2-y^2} = \sqrt{r^2\,(1-\sin^2\theta)}$, then $\sqrt{r^2\,\cos^2\theta}$ and $r\,|\cos\theta\,|$. Therefore

$$\int \sqrt{r^2-y^2}\, dy = \int r^2\,\cos^2\theta\, d\theta$$

We now apply Carnot formulas and simplify:

$$r^2\int \frac{(1+\cos 2\theta)}{2}\, d\theta = \frac{r^2}{2}\int (1+\cos 2\theta)\, d\theta$$

This is integrated as $\frac{1}{2}r^2\,(\theta + \frac{1}{2}\sin 2\theta) + C$. The identity $\sin 2\theta = 2\sin\theta\cos\theta$ transforms this to $\frac{1}{2}r^2\,(\theta + \sin\theta\cos\theta) + C$. We distribute $\frac{1}{2}r^2\,(\theta) + \frac{1}{2}r^2\,(\sin\theta\cos\theta) + C$ and return to the starting variable y, $\frac{1}{2}r^2\,\arcsin(y/r) + \frac{1}{2}r^2\,(y/r)\cos\theta + C$. We then apply $\cos\theta = \sqrt{1-\sin^2\theta}$ and simplify to $(1/2)r^2\,\arcsin(y/r) + (y/2)\sqrt{r^2-y^2} + C$. Member (1) will be twice this value,

$$(1) \;=\; 2\int_0^{R\frac{\sqrt{63}}{8}} \sqrt{r^2-y^2}\, dy \;=\; \left[r^2\,\arcsin\left(\frac{y}{r}\right) + y\sqrt{r^2-y^2}\,\right]_0^{R\frac{\sqrt{63}}{8}}$$

Since $F(0) = 0$, we obtain

$$16R^2 \arcsin\left(\frac{R\frac{\sqrt{63}}{8}}{4R}\right) + R\frac{\sqrt{63}}{8}\sqrt{16R^2 - \frac{63}{64}R^2}$$

By simplification, this reduces to

$$(1) \;=\; R^2\left[16\arcsin\left(\frac{\sqrt{63}}{32}\right) + \frac{31\sqrt{63}}{64}\right]$$

- We proceed in the same manner with (3):

$$(3) \;=\; 2\int_0^{R\frac{\sqrt{63}}{8}} \sqrt{R^2 - y^2}\,dy \;=\; \left[R^2 \arcsin\left(\frac{y}{R}\right) + y\sqrt{R^2 - y^2}\right]_0^{R\frac{\sqrt{63}}{8}}$$

Again $F(0) = 0$, so we get

$$R^2 \arcsin\left(\frac{R\frac{\sqrt{63}}{8}}{R}\right) + R\frac{\sqrt{63}}{8}\sqrt{R^2 - \frac{63}{64}R^2}$$

By simplification, this reduces to

$$(3) \;=\; R^2\left[\arcsin\left(\frac{\sqrt{63}}{8}\right) + \frac{\sqrt{63}}{64}\right]$$

The area of the lens is the sum of (1), (2) and (3):

$$\mathscr{A} \;=\; R^2\left[16\arcsin\left(\frac{\sqrt{63}}{32}\right) + \frac{31\sqrt{63}}{64}\right] - \sqrt{63}R^2 + R^2\left[\arcsin\left(\frac{\sqrt{63}}{8}\right) + \frac{\sqrt{63}}{64}\right]$$

A numerical estimate provides

$$\mathscr{A} \;=\; R^2\,(7.855 - 7.937 + 1.569)$$

Accordingly, the intensity of light at point E, which corresponds to the number of circles that lit point E, and whose centers are located in the lens-shaped region ABE, equals the area of this region:

$$I_E = \mathscr{A} = 0.473\,\pi R^2$$

3. Tip. Finally, when the scanning point arrives at point T on the tip of the crescent, it is lit by the circular lights whose centers are in the horn-shaped region ATB. There are as many patches of lights illuminating point T as there are points in the area of figure ATB (Fig. 4.25).

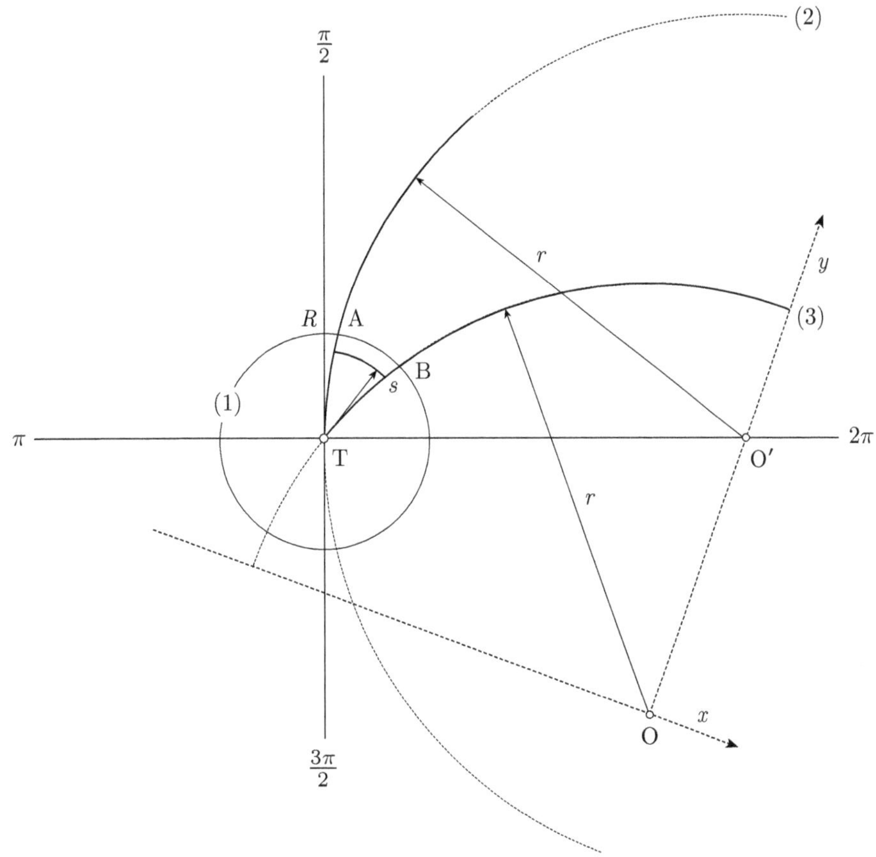

Fig. 4.25. The Amount of Light at the Tip of the Crescent

Although the horn-shaped region ATB is limited by three curves, this area is rectifiable, as we can see by switching to polar coordinates (θ, r). Let us place the origin in T and rotate the axes to make the equations as simple as possible. The equation of circle (1) is $r = R$, that of circle (2) is $r = 2r\cos\theta$, and that of circle (3) is

$$r = 2r\cos\left(\theta - \frac{2\pi}{9}\right)$$

This equation means that TA is equivalent to TB by a rotation of $\frac{2\pi}{9}$ around the origin. Angle $\frac{2\pi}{9}$ is independent of r and thus constant. Consequently, the area of the horn-shaped region ATB is simply equal to the area of the circular sector ATB. Therefore this area is written as

$$\mathscr{A} = \int_0^R \frac{2\pi}{9} r\, dr = \left[\frac{\pi}{9}R^2\right]_0^R = \frac{\pi}{9}R^2$$

As a result, the intensity of light at the tip of the crescent-shaped image is

$I_T = \mathscr{A} = 0.111\,\pi R^2$

We have now finished calculating the intensity of light when the point scans the central region, edge and tip of the entire crescent-shaped image.[30] The three situations indicate a wide variation in the intensity of light:

$$I_C = \pi R^2 \; > \; I_E = 0.473\,\pi R^2 \; > \; I_T = 0.111\,\pi R^2$$

When the projection plane is moved back and forth, the patch of light widens or shrinks, while keeping the same amount of light. When the plane is moved away, the patch of light widens and the center, edge and tip of the crescent lose emittance in proportion to their intensity. Accordingly, there comes some point where the intensity of light falls below the eye sensitivity threshold. At the very beginning, the tips vanish, then the edges, then the central region of the crescent-shaped figure. Therefore, Ibn al-Haytham was basically correct in his sketch of proto-photometry: the unequal intensity of light between the tips, edges and center is an additional factor to be taken into consideration in the explanation of the rounding of the image.

That said, it is hard to understand why Ibn al-Haytham claimed that the sickle moon was not seen through the camera obscura. In addition, other observers testified the revese, such as Egidius de Baisiu:

30. The limb darkening is another effect contributing to the vanishing of the edge. This effect was first described by Luca Valerio with the help of a darkroom (1612). Contested by Galileo (1613) but then confirmed by Scheiner (1630), it gave rise to measurements from Bouguer (1760) to Very (1902) and was finally explained by Schwarzschild (1906). The limb darkening is a density and temperature-driven effect. It makes the solar disc appear brighter on the central region. It is often written as $I_\theta = I_0[1 - u(1 - cos\theta)]$, where I_0 denotes the maximum intensity of light, θ is the angle between a line normal to the stellar surface and the line of sight, and u is a darkening coefficient. It has been demonstrated that the limb darkening varies in time as well as in function of the wavelength (Petro et al. 1984; Hestroffer and Magnan 1998).

> "Et videbis in pariete opposito foramini lumen in tali portione qualis est portio corporis luminosi ... Et similiter est de luna sive foramen sit rotundum sive non" (*Improbatio*, ed. Mancha 1989: 5–6).

and Gersonides:

> "It is best for the hole of the window to be very small, for then the rays that arrive at the wall that receives the light take on the shape of the Moon according to the amount of the eclipse" (*Astronomy* V.22, ed. Goldstein 1985: 49).

Our observations of the waxing crescent with a 2-meter helioscope equipped with a 10-mm aperture also confirmed the visibility of the lunar crescent on the projection plane. There are several possible explanations for this discrepancy.

1. *Authoritative argument*: Ibn al-Haytham could have revived the opinion of the *Problemata Physica*, without observation. This is unlikely because Pseudo-Aristotle's opinion appears in Chapter XV, 11, which is lacking in Ḥunayn ibn Isḥāq's version.

2. *Atmospheric extinction*: Ibn al-Haytham could have observed the Moon either when it was low above the horizon, or at a time when the air was dust laden. Either factor could have resulted in attenuating the lunar brightness.

3. *Small aperture*: Ibn al-Haytham may also not have observed the crescent moon when the dark chamber was equipped with a tiny aperture, so that the light received on the projection plane was below the sensitivity threshold.

From a distance of a millennium, and having no precise knowledge of the viewing conditions available to Ibn al-Haytham, it is virtually impossible to decide between these explanations. Historical progress can only be expected on the first explanation.

Conclusion

Having studied the scientific sources, optical methods and results achieved by Ibn al-Haytham in his treatise *On the Shape of the Eclipse*, we are now in position to take a step back and assess his pioneering work as a whole. In spite of a few limitations—the lack of clarity of certain choices, such as the factor ten involved in the definition of a "circular image," or the failure to draw a line between proto-photometric and geometric considerations, which is understandable given the historical context—Ibn al-Haytham's work as a whole provided new and robust knowledge.

This was new knowledge, because Ibn al-Haytham did not limit himself to repeat the camera obscura of his forerunners nor did he reproduce the conventional observations conducted with it. Of foremost importance, testing small apertures allowed him to produce an optical image in the strictest sense of the word—something that had never been done before. In the worldwide history of optics, *On the Shape of the Eclipse* is both the first mathematical work on the operating of the camera obscura and an unprecedented study of the conditions of formation of images, as defined in the narrow optical sense of the word.

This was robust knowledge, because Ibn al-Haytham freed himself from the context in which the camera obscura was commonly used (that of verifying the linear propagation of light, dealt with in *Optics* 1, 3) to engage in a thorough understanding of its operating. He achieved this goal through a combination of mathematics and the experimental method, which was no less efficient in medieval Islam than in the modern era. The most striking feature is the systematic way Ibn al-Haytham conducted the experiments: in *On the Shape of the Eclipse*, he varied all that could be changed: the shape and size of the aperture, the focal length of the dark chamber, the distance and shape of the celestial bodies.

Full knowledge was derived from this experimental set-up.

Appendix[31]
A Tentative Dating of Ibn al-Haytham's *On the Shape of the Eclipse*

This Appendix is devoted to the astronomical dating of Ibn al-Haytham's *On the Shape of the Eclipse*. After clarifying the conditions of observation, an astronomical ephemeris is applied to sift through the solar eclipses that occurred during Ibn al-Haytham's life throughout the area he is believed to have sojourned. Next, the remaining eclipses are sorted by computing the image projection. The results show that Ibn al-Haytham's work is likely to have reported the partial solar eclipse of 28 Rajab A.H. 380/21 October 990 in Baṣra. This finding, however, should be regarded as a tentative outcome, pending further information.

1. The Status of Scientific Diagrams

Scientific texts usually include diagrams. In the physical sciences, these diagrams have a special status because they are expected to represent real physical bodies. These analog representations exhibit structural similarities with the visual percept. Little is known about the codes of representation used in texts that originated in ancient times. In particular, it is unclear whether diagrams faithfully reproduced the physical bodies they depicted. It is frequently assumed that they did not; instead they may have represented the physical bodies in a schematic form. Historians of science generally consider that diagrams transmit untrustworthy knowledge, regarding the morphological characteristics of the objects depicted.

31. This section is adapted from "A Tentative Astronomical Dating of Ibn al-Haytham's Solar Eclipse Record," *Nuncius* 29 (2014): 324–358.

D. Raynaud, *A Critical Edition of Ibn al-Haytham's On the Shape of the Eclipse*,
Sources and Studies in the History of Mathematics and Physical Sciences,
DOI 10.1007/978-3-319-47991-0

Although it may be convenient, the reasoning described above is not well-grounded. Given the premise of a lack of knowledge of how the diagrams were drawn, scholars must conclude that either of the following hypotheses is possible: The diagrams were analog representations and were faithful to reality (H1), or they were not (H2). In mathematics, H2 is a leading hypothesis, because geometric diagrams are often simplified and over-specified (Saito 2005; Sidoli and Saito 2009). In astronomy, H1 prevails even for ancient sky charts. It has been proven that the positional accuracy of the Dunhuang chart is of the order of 1.5° to 4° (Bonnet-Bidaud, Praderie and Whitfield 2009). Similar conclusions could be drawn from al-Ṣūfī's *Catalogue of Fixed Stars* (*Ṣuwar al-kawākib al-thābita*). A test based on the Orion constellation shows that al-Ṣūfī's chart is accurate from 1° to 3°.[32] These documents are reasonably faithful to reality.

When a work mathematically analyzes a natural phenomenon, it stands at the junction of these two trends—the abstract and the realistic. Thus, the choice between H1 and H2 is trickier. Truthfulness of a diagram is the most probative when (a) the depicted object has a simple form, (b) careful observations are documented, (c) the work deals with optics and vision, and (d) the text records real observations. Ibn al-Haytham's work meets all four of these conditions (a, b, c, and d).

(a) The shape of a partial solar eclipse is a simple figure, reduced to the intersection of two circles.

(b) There is a long tradition of measuring eclipses. Arab scholars performed astronomical observations with great accuracy (Stephenson and Said 1991). To illustrate the point, consider what al-Mahānī says about a partial eclipse of the Sun that he observed in Baghdād on 16 June 866: "Eclipse began at 6h03m, unequal hours, mid-eclipse at 7h10mn, the end at 8h16m. Duration of the eclipse 2h16m. Size on the solar diameter $9\frac{1}{12}$ digits, corresponding to 8 digits of the surface. The apparent place of the Sun at mid-eclipse 23°29′ Gemini, the place of the Moon at the same time 23°47′ Gemini." (Ibn

32. The test was conducted by superimposing the standard IAU sky chart onto the corresponding page of the Bodleian, MS Marsh 144, dated 1009 CE, where each constellation is projected onto the plane tangent to the sphere. In the Orion constellation, among 32 stars, 17 fall within one degree of their actual position ($\alpha, \beta, \gamma, \delta, \varepsilon, \zeta, \varkappa, \lambda, \mu, \nu, \pi^2, \pi^3, \pi^4, \pi^5, \varphi^1, \xi, \psi$), 10 within two degrees ($\vartheta, \iota, o^2, \pi^1, \sigma, \tau, \chi^2, \varphi^2, \upsilon, \omega$), and 3 within three degrees (χ^1, ρ, η), two stars are missing (o^1, π^6). For general comments on al-Ṣūfī's star catalogue (Grasshoff 1990: 20-21; Goldstein and Hon 2007).

Yūnus 1804: 92). Ninth-century reports adhered to a strict protocol, indicating (a) first, maximum, and last contact; (b) an eclipse magnitude expressed in digits (*aṣābiʿ*), a "digit" being one-twelfth of the solar diameter; and (c) an eclipse obscuration expressed in equaled digits, one "equaled digit" being one-twelfth of the solar disc area. Even when the astronomer could not perform a measurement in due form, he would give an estimate.[33]

(c) The treatise has four diagrams. Diagram 1 shows the eclipse in the sky and on the projection plane. Diagram 3 is a simple variant of Diagram 1. The other two diagrams are portions of Diagram 1. The work as a whole is a comment on a single diagram, which is central for understanding the significance of a work that envisions the solar eclipse as an optical phenomenon. Additionally, in the *Optics*, III, 7, Ibn al-Haytham provided an in-depth consideration of realistic depictions. He especially stated that "pictures look like the visible bodies to which they correspond... paying particular attention to points of resemblance." (Sabra 1989, I: 295). Hence, his concern about representational issues.

(d) Ibn al-Haytham genuinely observed a partial solar eclipse. His text has several items of *muʿtabir* (experimenter) and *iʿtabara* (to experiment, to test by confronting observations). Ibn al-Haytham writes: (1) "The aperture, by which the sun[light] is tested..." (2) "The diameter of the aperture, by which the sun is tested..." (3) "When one tests/experiments sunbeams issuing from a wide aperture..."[34]

These conditions prove that Ibn al-Haytham actually tested the crescent-shaped image of a partial solar eclipse, and did not leave the diagrams to chance. One can be confident that the diagrams of *On the Shape of the Eclipse* are truthful.

However, the time and place of the eclipse can be determined by astronomical methods only if the position of the image in the darkroom is known. In the *Optics*, for example, the sunlight is said to reach "the chamber's floor or walls" (Sabra 1989, I: 13). Therefore it is unclear whether, in the case of the observation of the solar

33. Al-Bīrūnī reported observing a solar eclipse on [8] April 1019 while he was traveling: "We were near Lamghān, between Qandahār and Kābul, in a valley surrounded by mountains, where the Sun could not be seen unless it was at an appreciable altitude above the horizon. At sunrise, we saw that approximately one third of the Sun was eclipsed and that the eclipse was waning" (1967: 261).

34. Arabic has three conditional forms: eventuality (*iḏā*), potentiality (*in*), and irreality (*law*). In Ibn al-Haytham's text, the verb *iʿtabara* is always coupled with the indicative (5 times), as in sentences 1-2 (lines 670–1), or with the first form of the conditional (twice), as in sentence 3 (lines 28–9):

(١) الثقب الذي يعتبر به الشمس ... (٢) قطر الثقب الذي يعتبر به الشمس ... (٣) وإذا اعتبر أضواء الشمس التي تخرج من الثقب ...

eclipse, the image of the Sun was cast onto the floor or onto the wall. The solution can be derived from astronomical data. During a solar eclipse, the altitude of the Sun varies continuously from Alt_B (beginning) and Alt_m (maximum) to Alt_E (end). Let H be the number of "horizon eclipses," such that $0° \leqslant Alt_m \leqslant 10°$, and Z be the number of "zenithal eclipses," such that $80° \leqslant Alt_m \leqslant 90°$. These are counted in Tables App3–6 (pp. 183–6):

H	Z	Alt_m	$Year$	$Place$
10	0	70°	0993	Baṣra
12	0	72°	1037	Cairo
12	0	72°	1007	Damascus
11	0	77°	1007	Baghdād

It appears that no zenithal eclipse occurred during the whole period 970–1038 (column 2). Assuming that the image was cast onto the floor, then an altitude $70° \leqslant Alt_m \leqslant 77°$ (column 3) would cause a deformation of the image, contrary to the mention of circles—not ellipses—in the treatise. Since there were 10 to 12 horizon eclipses in the same period (column 1), the comparison reveals that the projection plane (*al-saṭḥ muwāzī li-saṭḥ al-thaqb*), on which the crescent-shaped image was inspected, was in a vertical position.

Consequently, the time and place of the eclipse can be determined by astronomical methods (Steele 2000; Stephenson and Steele 2006). In what follows, astronomical dating is combined with the computation of the image projection.

2. The Eclipses to be Surveyed

The observation conditions have been defined clearly enough to allow the consideration that Ibn al-Haytham's drawings were faithful to reality. Therefore, the partial solar eclipse can be dated by astronomical methods through the analysis of the solar eclipses visible throughout Ibn al-Haytham's lifetime.

Dates. Since Ibn al-Haytham died in or shortly after 1040, there is no need to look at events that occurred before A.H. 359/970, a date allowing for a 70-year-long career. The year 970, then, can serve as the *terminus post quem* of the period to be surveyed. Since *On the Shape of the Eclipse* appears in the so-called "List III"—a register of Ibn al-Haytham's works which dates from 2 October 1038—this date can serve as the *terminus ante quem*.

Places. Ibn al-Haytham was born and educated in Baṣra.[35] Today, the ancient city of Baṣra is located in Zubayr, 20 km south-west of Baṣra. Ibn al-Haytham moved to Cairo during the caliphate of al-Ḥākim. The bio-bibliographers disagree about what he did next: (a) according to Ibn al-Qifṭī (1903: 165), he remained in Cairo after the Nile episode; (b) according to al-Bayhaqī (1935: 77–80), he fled to Damascus where he found a well-to-do patron; (c) others think that he returned to ʿIrāq (Ibn Abī Uṣaybiʿa 1882-4, II: 90). If Ibn al-Haytham ever visited Baghdād, it did not occur before 987, because it is not mentioned in the *Fihrist* (Ibn al-Nadīm 1970). Ibn al-Haytham's presence in Cairo is poorly documented. Ibn Abī Uṣaybiʿa tell us that Isḥāq Ibn Yūnus took down some notes on Diophantus from Ibn al-Haytham in Egypt (date unknown), and al-Andalusī (1912: 60) notes an encounter between Ibn al-Haytham and the qāḍī of Toledo in Egypt in A.H. 430/1038.

Since the sources contradict each other, all possible locations must be considered:

Baṣra	*Cairo*	*Damascus*	*Baghdād*
30° 23′ 29″ N	30° 02′ 45″ N	33° 30′ 42″ N	33° 20′ 30″ N
47° 42′ 20″ E	31° 15′ 44″ E	36° 18′ 24″ E	44° 23′ 07″ E
Elevation 0 m	Elevation 33 m	Elevation 698 m	Elevation 40 m

Eclipses witnessed during the period 970–1038 may be found by entering coordinates in an ephemeris. I used the NASA Eclipse Predictor, which is based on the solar system theory VSOP87 (Bretagnon and Francou 1988) and the lunar theory ELP2000-82B (Chapront-Touzé and Chapront 1983).[36] In the half century extending

35. There were two cities called Baṣra in medieval Islam: one in ʿIrāq, and the other—called Baṣra al-Ḥamra—in Morocco (Eustache 1955). There are several indications that Ibn al-Haytham lived in Baṣra, ʿIrāq; for example, the many words of Persian origin found in his texts.

36. Theory ELP2000-82B ensures an accuracy of 20″ for the Earth-Moon couple. In the NASA program, very small periodic terms—with coefficients smaller than 0.0005″ in longitude and latitude—are ignored. As a result, the Moon's position has a mean error of about 0.0006″ of time in right

from 970 to 1038, 30 partial solar eclipses occurred, in a wide range of positions and magnitudes (Tables App3–6, pp. 183–6).

3. The Magnitude of the Eclipse

The projected crescent-shaped images—that is, the three superimposed lunes in Figs. 5–9—show little deviation from one diagram to another. This ensures that the handwritten tradition preserved an essential feature of Ibn al-Haytham's text. The magnitude is measured from the manuscripts and then compared to the theoretical magnitude given by the ephemeris.

Each of the five manuscripts has two crescent-shaped diagrams, that is, Diagrams 1 and 3. Each diagram provides one image in the sky and three projected images, totaling 50 interlocking circles. Given their large number, these figures allow for statistical treatment.[37] The magnitudes computed from the dimensions measured on the manuscript diagrams range from $0.623 \leqslant m \leqslant 0.872$, with mean $\mu = 0.779$—that is about 9 ¼ digits—and standard deviation $\sigma_1 = 0.0568$.

---

ascension and 0.006″ in declination. This uncertainty accounts for approximately one percent of the Sun's diameter. Uncertainty in the Moon's secular acceleration is slightly higher. It has been estimated in 1000 CE to be about $\Delta T = +1570 \pm 54''$.

37. Since the distances and diameters of the Sun and the Moon cannot be in proportion, the scribes solved the problem in three different ways (Figure 1, dotted line): (1) MSS P and F depicted the sky image in accordance with the magnitude by moving the center Ṣ (peak 0.787); (2) MS B put the centers S and Ṣ in the right places by altering the magnitude of the sky image (peak 0.500); (3) MSS O and L adopted a middle way (peak 0.625). This is why I reject the sky images and proceed with the projected images only. I used an unbiased estimator of standard deviation, which is convenient for small samples:

$$\sigma_1 = \frac{s}{c_4} \text{ with } s = \sqrt{\frac{1}{n-1}\sum_{i=1}^{n}(m_i - \mu)^2} \; ; \; c_4 = \sqrt{\frac{2}{n-1}} \cdot \frac{\Gamma(n/2)}{\Gamma((n-1)/2)} = 1 - \frac{1}{4n} - \frac{7}{32n^2} - \frac{19}{128n^3} = 0.991$$

where n is the sample size and Γ the gamma function $\Gamma(n) = (n-1)!$ Estimator σ_1 has been applied to magnitudes m_3 ... m_5 and m_8 ... m_{10} as measured on the manuscripts. Magnitudes: m_1 crescent ABJD, m_2 crescent A(Ṣ)J, m_3 crescent ŠYF, m_4 crescent KLMN, m_5 crescent ṮẒGQ (on Diagram 1); m_6 crescent ABJD, m_7 crescent A(Ṣ)J, m_8 crescent ŠYH̱F, m_9 crescent KLMN, m_{10} crescent ṮẒGQ (on Diagram 3; Plate App6, p. 261).

These values are close to the ones provided by MS P: $\mu = 0.749$, $\sigma_1 = 0.0368$.[38] Because of the consistency of all its diagrams, MS P may be seen as the best diagrammatic witness of *On the Shape of the Eclipse*, a conclusion that accords with the stemmatological analysis of this treatise (Raynaud 2014*c*).

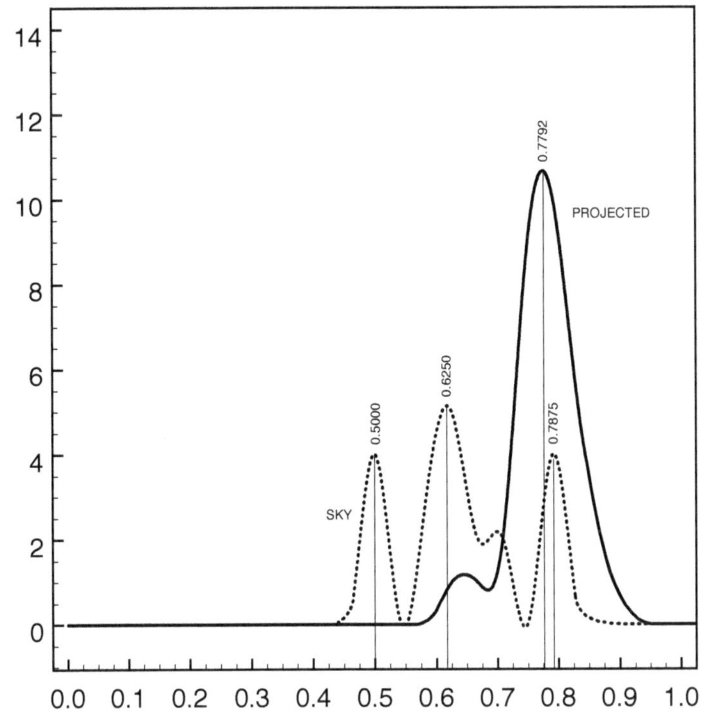

Fig. App1. The Magnitude of the Eclipse

38. The dimensions measured on the manuscripts are the following:

MS	m_1/m_6	m_2/m_7	m_3/m_8	m_4/m_9	m_5/m_{10}
F	43.0/54.5=0.789	43.0/54.0=0.796	51.0/66.5=0.767	48.5/63.5=0.764	57.0/71.0=0.803
	41.5/65.0=0.638	41.5/65.0=0.638	50.0/63.0=0.794	50.0/62.0=0.806	50.0/62.5=0.800
B	43.3/85.7=0.505	43.3/86.9=0.498	73.7/87.6=0.841	52.8/67.0=0.788	53.2/65.5=0.812
	42.6/84.3=0.505	42.6/84.3=0.505	37.3/53.0=0.704	43.0/53.2=0.808	45.5/52.2=0.872
P	39.3/50.0=0.786	39.3/50.0=0.786	59.8/79.0=0.757	58.9/79.4=0.742	66.5/79.0=0.842
	36.0/55.2=0.652	36.0/55.0=0.654	50.6/66.0=0.767	49.8/66.7=0.747	50.3/66.2=0.760
O	60.0/86.8=0.691	60.0/86.8=0.691	69.2/94.0=0.736	68.0/92.0=0.739	71.2/91.0=0.782
	54.5/90.0=0.605	54.5/90.2=0.604	66.0/84.2=0.784	74.0/85.3=0.867	68.2/85.5=0.798
L	42.0/66.2=0.634	42.0/66.0=0.634	41.0/49.0=0.837	44.5/52.8=0.843	40.0/48.8=0.820
	45.8/77.1=0.594	45.8/77.1=0.594	24.2/35.7=0.678	24.5/35.2=0.696	21.8/35.0=0.623

After eliminating the solar partial eclipses whose magnitudes were too far from the mean, the remaining data encompassed two-thirds of the series of events: $\mu - \sigma_1 = 0.722 \leqslant m \leqslant \mu + \sigma = 0.836$. Solar eclipses fitting the manuscript values are marked with an asterisk (Tables App3–6 provide the complete set of data).

Date	*Baṣra*	*Cairo*	*Damascus*	*Baghdād*
0970 May 08	0.622	* 0.779	* 0.802	0.718
0990 Oct 21	* 0.744	0.687	* 0.774	* 0.793
1000 Apr 07	0.891	0.575	0.605	* 0.760
1015 Jun 19	0.947	* 0.786	* 0.768	0.853
1030 Aug 31	0.633	0.660	* 0.748	* 0.769
1032 Jan 15	* 0.813	0.622	0.600	0.699
1033 Jun 29	0.839	0.645	* 0.791	0.871

The removal of Cairo 1037 because of its low magnitude ($m = 0.058$) is also justified on historical grounds. List III of Ibn al-Haytham's scientific works is known in three versions, and the versions of Lahore and Kuibyshev differ from the one transmitted by Ibn Abī Uṣaybiʿa. The list of Ibn al-Haytham's works that were discovered (Rozenfeld 1975, 1976) in a manuscript in the V.I. Lenin Kuibyshev Regional Library bears the explicit note: "Fihrist taṣānīf al-Ḥusayn ibn al-Ḥasan ibn al-Haytham li-sanat 427", i.e., "Catalogue of the works of al-Husayn [i.e., al-Ḥasan] ibn al-Ḥasan ibn al-Haytham up to the year 427" (ending 24 October 1036). The Kuibyshev Library thus holds an intermediate list of works copied from the same source before Ibn Abī Uṣaybiʿa reproduced it. Since *On the Shape of the Eclipse* appears in Kuibyshev's list, it must have been written before 24 October 1036.

4. The Occultation Angle

Among the five manuscripts of Ibn al-Haytham's work, MSS B and L present the eclipse diagrams vertically, and MSS F and O present the diagrams horizontally. The first figure of MS P is vertical, but its lettering runs horizontally, like MSS F and O. Therefore, we must decide between the two sets of manuscripts based on the text alone.

Luckily, the positions of the Sun and the Moon are stipulated in the text itself (compare the following text to Plate App6, p. 261):

> "Line FN becomes smaller than the semi-diameter of arc KNM... Then circle ŠṮ cuts the line KF *below* (*taḥta*) the arc ŠF."[a] "Circumferences [of all circles whose radius is ŠṮ, and whose centers are taken onto KNM] go just *below* (*taḥta*) point F."[b] "Line NF is larger than the semi-diameter of arc KNM, thus circle ŠT cuts the line KF *above* (*fawqa*) point F."[c][39]

Whatever the diagram's orientation, all the manuscripts have the same text.[40] As the center of arc KNM is below F, the convex part of the crescent-shaped image (i.e., the Sun's limb) is up. Therefore, the lectio of MSS B and L, which were preferred by Wiedemann, must be withdrawn in favor of MSS F, P and O, which were adopted by Naẓīf and Sabra. These MSS testify that the Sun was above the Moon.

This being established, an ephemeris was employed to search for the position of the Sun and Moon at the time of the eclipse. I used the NASA-based Horizons program, which is built upon the planetary and lunar theories LE/DE200-406.[41] Body center coordinates *Alt, Az* are written in 0.0001 degrees, including corrections for light-time, the gravitational deflection of light, stellar aberration, precession, nutation and atmospheric refraction.

Δ *Alt*, Δ*Az*. *Altazimuthal distances between the centers of the Sun and Moon.* These distances are calculated directly from object coordinates.

---

39. Lines 492–9:
a فيكون خط نف أصغر من نصف قطر قوس كنم ... فدائرة شث يقطع خط كف تحت قوس شف. b فجميع هذه الدوائر (الدائرة F) يكون محيطاتها تحت نقطة ف. c فيكون خط نف أعظم من نصف قطر قوس كنم ... ويكون دائرة شث تقطع خط كف فوق نقطه ف.

40. These words appear in MS F 60r:25, 60v:1, 60v:6, 60v:12 / MS B 95r:8, 95r:12, 95r:20, 95v:8 / MS P 40v:9, 40v:14, 41r:4, 41r:17 / MS O 84r:21, 84r:23, 84r:28, 84r:35 / MS L 25v:11, 26r:2, 26r:8, 26v:4 for *taḥta* (thrice) and *fawqa* (once).

41. The JPL-NASA model does not undertake the correction of 0.25″ in latitude and 0.50″ in longitude of the Moon's position. Two Lunar radii (k1 = 0.272 4880 and k2 = 0.272 2810) are used instead of the one value recommended by the UIA (k = 0.272 5076). Both choices affect the eclipse magnitude. I do not proceed with the corrections because the differences are negligible in comparison to the Sun's diameter.

γ. *Occultation angle.* This is the angle standing at the center of the Sun, between the vertical line and the axis that joins the centers of the Sun and Moon (C_S and C_M). The angle is counted clockwise from the top. Any angle value depends on the sign of ΔAlt.

$$\gamma = \arctan\left(\frac{\Delta\, Az}{\Delta\, Alt}\right) + k\pi \qquad \begin{cases} k = 0 & \text{if } \Delta\, Alt + \\ k = 1 & \text{if } \Delta\, Alt - \end{cases}$$

In the manuscripts, axis ṢB, which joins the centers of the Sun and Moon, is rarely parallel to the axis ẒT, which links the centers of the projected images. This may indicate that Ibn al-Haytham was representing the angle of occultation. The greatest angle appears in the most reliable manuscript: 8° 00′ on MS P. Therefore, the occultation angle could have been higher than the mean angle $\bar{\gamma} = +1°54'$.[42] In the absence of additional information, I will simply place an asterisk next to the data $\bar{\gamma} - 45° \leqslant \gamma \leqslant \bar{\gamma} + 45°$ and eliminate all the events that do not fit.

Date	*Baṣra*	*Cairo*	*Damascus*	*Baghdād*
0990 Oct 21	* 32.964	* 8.726	* 12.146	* 25.099
1030 Aug 31	82.361	* 38.896	* 36.302	* 38.073
1037 Apr 18		* 19.284	58.822	

The combination of criteria 5 (magnitude) and 6 (occultation angle) reveals that Ibn al-Haytham could have reported only two eclipses: either that of 21 October 990 (Baṣra, Damascus or Baghdād) or that of 31 August 1030 (Damascus or Baghdād).[43] Other conditions can help identify the proper dates and places. The observation of an eclipse must be compatible with the Sun's position in the sky. Therefore, I remove all places where the horizon would have been obstructed at sunrise or sunset.

42. Compare: MS F 121r: 0° 00′, 122v: +4° 30′ / MS B 88v: –2° 00′, 96r: +1° 00′ / MS P 31v: +8° 00′, 42v: +2° 00′ / MS O 81r: –0° 30′, 84v: +3° 00′ / MS L 34r: +1° 30′, 34r: +1° 30′.

43. The difference in the angular diameters of the Sun and the Moon cannot act as a criterium: "As shown by the mathematicians," Ibn al-Haytham states, "the cone bounding the Sun sphere is equal to the cone bounding the Moon sphere [...] and the crescent-shaped figure is bounded by two arcs of two equal circles." Lines 122–5:

... شكل الهلالي يحيط به قوسان من دائرتين متساويتين.

Damascus 1030. The city is surrounded by mountains on the north-west. Azimuth $Az = 273°$ points to Jabal al-Shaykh with peaks 2,814 m high. Thus, the 1030 eclipse would not have been compatible with the site's geography. This conclusion again is in accordance with the historical documentation: Ibn al-Haytham does not appear once in the *History of Damascus* (Ibn ʿAsākir 2000, 75, s.v.).

Baghdād 1030. With $Alt = 1°$, the Sun was setting at the time of the eclipse. Thus its visibility would have been hampered by the buildings of the city. This location also raises two additional problems. First, the image would have been strongly distorted, as seen in the diagram above Table App6. Second, considering Rashed's finding that al-Ḥasan ibn al-Ḥasan and Muḥammad were two distinct persons, it is likely that Ibn al-Haytham never returned to ʿIrāq (Rashed 2000: 937–941). Another reading has been supported by Sabra (1998, 2002–3). Since then, however, Rashed added some other decisive arguments (Rashed 2007).

Consequently, only three possible locations—Baṣra, Damascus and Baghdād—were in fact compatible with the observation of the partial solar eclipse of 990 CE.

5. The Geometry of the Eclipse Image

As we noted earlier in Chapter 3.5, p. 99, the crescent-shaped image provided by the eclipse consisted of two interlocked circles—not ellipses. Consequently, all eclipses that would have caused a sharp distortion of the image must be eliminated.

Location Known

In a first scenario, the place from where the eclipse was observed is known. By considering its urban situation, I can calculate the vertical and horizontal distortions of the image.

Image distortion can be calculated from the conjugation between the object-points in the sky and corresponding image-points on the plane. This strategy creates a point-by-point image, however, which constitutes a major flaw when reconstructing the image from geometric elements.

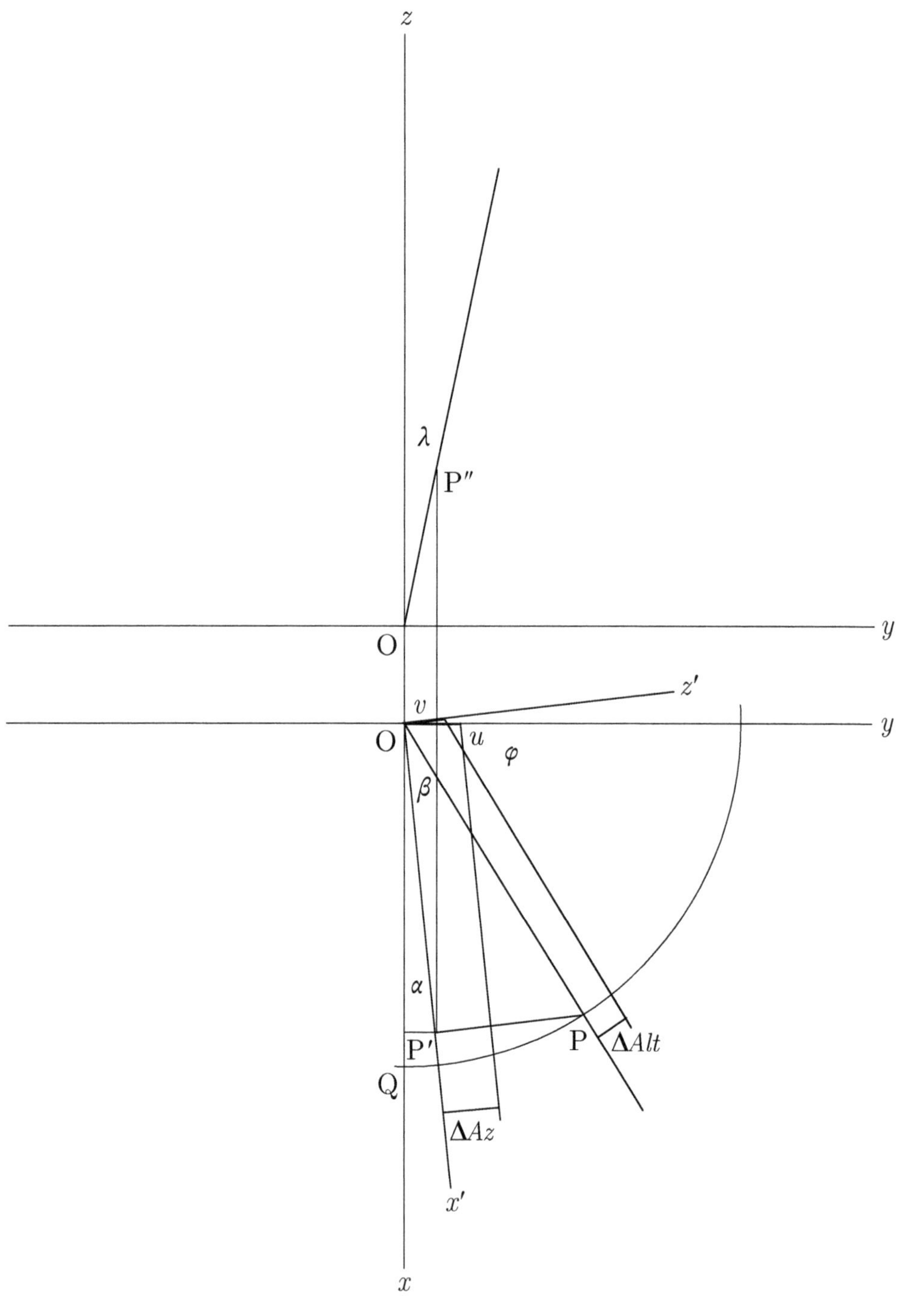

Fig. App2. Projection Elements ($\alpha, \beta, \varphi, \lambda$; u, v)

Consider that the ray falling onto the wall forms with the projection plane an angle $\varphi = \widehat{\mathrm{PO}z'}$ such that $0 \leqslant \varphi \leqslant \frac{\pi}{2}$ and $\varphi = \frac{\pi}{2}$ if the radius is normal to the plane. This is the easiest way to determine the deformation of the image cast onto the opposite wall of the darkroom. Whatever the case, the solar disc appeared as a conic section, most often as an ellipse. Angle φ is calculated from an altitude and azimuth (*Alt, Az*) embedded in an urban context, thus defining the relative altitude and azimuth (α, β) (Fig. App2).

α. *Relative Azimuth.* The azimuth of the light beam—that is, the angle relative to the vertical plane normal to the projection plane—must be determined. Since this angle depends on the darkroom's orientation, the matter will be discussed in due course (this Appendix, Section 6, p. 175).

β. *Relative Altitude.* The altitude of the light beam must be identified too. This is the angle between the light beam and the horizontal plane normal to the projection plane. $\beta = \widehat{x'\mathrm{OP}}$, or equivalently $\beta = Alt$.

I will now calculate, step-by-step, angle φ from α and β.

ϑ. *Light beam angle with the normal line to the projection plane.* Let (O, x, y, z) be the orthonormal basis, with the origin O marking the point of impact of the sun beam onto the projection plane $\mathscr{P}$. Point Q (1, 0, 0) defines a unit vector $\| \vec{q} \| = 1$ in the direction Ox normal to the projection plane. The coordinates of a point P, such that $\| \mathrm{OP} \| = \| \vec{p} \| = 1$, are:

$$\mathrm{P} \begin{pmatrix} \cos\alpha\ \cos\beta \\ \sin\alpha\ \cos\beta \\ \sin\beta \end{pmatrix}$$

Angle $\vartheta = \angle \vec{p} \cdot \vec{q}$ can be calculated directly from the scalar product of vectors $\vec{p} \,.\, \vec{q} = \| \vec{p} \| \, \| \vec{q} \| \cos(\varphi)$ vectors. If $\| \vec{p} \| = \| \vec{q} \| = 1$, then we get $\cos(\varphi) = \vec{p} \,.\, \vec{q}$ and $\vartheta = \arccos(\vec{p} \,.\, \vec{q})$. The scalar product $\vec{p} \,.\, \vec{q}$ is computed separately by means of the P and Q coordinates:

$$\vec{p} \,.\, \vec{q} = \begin{pmatrix} \cos\alpha\ \cos\beta \\ \sin\alpha\ \cos\beta \\ \sin\beta \end{pmatrix} . \begin{pmatrix} 1 \\ 0 \\ 0 \end{pmatrix} = \cos\alpha\ \cos\beta$$

Hence:

$$\vartheta = \arccos\,(\cos\alpha\,\cos\beta)$$

φ. *Minimum angle of the light beam to the projection plane.* Since vector $\vec{q}$ is normal to the projection plane, angle $\varphi = \widehat{\mathrm{PO}z'}$ is complementary to ϑ:

$$\varphi = \frac{\pi}{2} - \arccos\,(\cos\alpha\,\cos\beta)$$

Angle φ determines the flattening of the ellipse. Since the source is very distant, I will assume the apex angle of the light beam is $\delta \approx 0$ and will reduce it to a cylindrical beam.

The light beam makes an angle φ to the projection plane $\mathscr{P}$, and the trace of the cylinder is an ellipse with axes $2a$ and $2b$. The focal axis of the ellipse is based on the Sun's apparent diameter $2b$:

$$2a = \frac{2b}{\sin\,(\varphi)}$$

λ. *Ellipse focal axis orientation*, relative to the direction Oz, is given by the orthogonal projection of the solar radius. Unit vector $\vec{p}$ in the projection has length $t = \sqrt{(\sin\,\alpha\,\cos\,\beta)^2 + (\sin\,\beta)^2}$. Therefore, angle $\lambda = \widehat{z\mathrm{OP}''}$ is:

$$\lambda = \arccos\left(\frac{\sin\,\beta}{t}\right) = \arccos\left(\frac{\sin\,\beta}{\sqrt{(\sin\,\alpha\,\cos\,\beta)^2 + (\sin\,\beta)^2}}\right)$$

$(u,\ v)$. *The coordinates of the Moon's center.* The projection changes the relative position of the Moon's center, which was expressed, in the case of the sky image, by the paired coordinates $(\Delta Az,\ \Delta Alt)$. Let us assume that the ray emitted by the center of the Sun reaches the projection plane at point O, the origin of (O, y, z). Taking into consideration image reversion, these are the coordinates $(u,\ v)$ of the Moon's center:

$$u = \frac{\Delta Az}{\cos\,(\alpha)} \quad \text{and} \quad v = \frac{\Delta Alt}{\cos\,(\beta)}$$

The geometric elements needed to build the image in projection are known. Fig. App2 provides a visualization of these data. Numeric values appear below.

Location Unknown

In a second scenario, we do not know the place from where the eclipse could have been observed. Without information regarding where the darkroom may have been located in the city, the horizontal distortion of the image cannot be calculated. The study is reduced to minimal vertical distortion when—by hypothesis—the Sun's azimuth was normal to the projection plane. Hence, the actual distortion was greater than this one. The calculus is simpler than if the location is known:

$$\varphi = \frac{\pi}{2} - Alt \qquad\qquad 2a = \frac{2b}{\sin(\varphi)}$$

6. Images in Vertical Distortion

The geometric elements previously defined, pp. 168–70 and 173–5, are computed (Table App1) and the eclipses observed in Baṣra, Damascus and Baghdād are drawn (Fig. App3). I consider Cairo 990 ($m = 0.687$) as an additional hypothesis.

	$Basṛa_{990}$	$Cairo_{990}$	$Damascus_{990}$	$Baghdād_{990}$
UT	11:47:54	11:14:21	11:20:40	11:37:16
ΔT[a]	± 56.5	± 56.5	± 56.5	± 56.5
Az_S	234.766	211.423	217.255	228.699
Alt_S	26.423	41.659	35.544	28.750
Az_M	234.832	211.445	217.276	228.739
Alt_M	26.524	41.807	35.644	28.834
ΔAz	+3.930	+1.362	+1.290	+2.358
ΔAlt	+6.060	+8.874	+5.994	+5.034
$2b_S$	32.331	32.331	32.331	32.331
$2b_M$	29.782	29.888	29.849	29.800
ϑ_S	26.423	41.659	35.544	28.750
φ_S	63.577	48.341	54.456	61.250
$2a_S$	36.103	43.275	39.735	36.877
v_M	+6.767	11.878	7.367	+5.742

Table App1. UT in h:mn:sec, ΔT in sec. Other data expressed in decimal degrees, except ΔAz, ΔAlt, $2b$, $2a$, and v presented in arcmns. All data calculated from Giorgini, NASA Jet Propulsion Laboratory, except [a] Uncertainty ΔT interpolated from tables created by Espenak and Meeus (2006, 13). Lunar data calculated similarly.

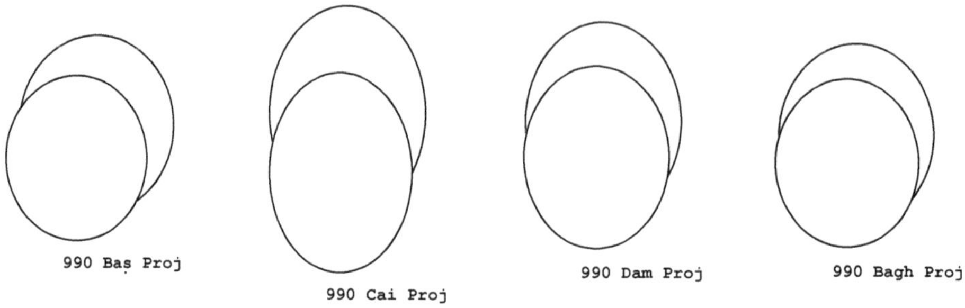

Fig. App3. Eclipse Images in Vertical Distortion

The eclipses seen in Cairo and Damascus must be removed because of strong vertical distortion ($Alt_{\text{Cai}} = 41°$; $Alt_{\text{Dam}} = 36°$). As we will see below, this removal is consistent with historical data.

Damascus 990. As explained in this Appendix, Section 4, p. 171, there is no mention of Ibn al-Haytham in Damascus, and most science historians have doubted that he ever sojourned in that city. His escape to Syria is putative, since it was only narrated by al-Bayhaqī (Meyerhof 1948: 155; Sabra 1989, I: xxxi; Rashed 1993: 4).

Cairo 990. Ibn Yūnus, an astronomer in Cairo, reported some 55 astronomical observations between 977 and 1007, but these do not include the solar eclipse of 21 October 990 (Ibn Yūnus 1804: 172–4). Furthermore, there is no documentary evidence of any early presence of Ibn al-Haytham in Cairo. It is well known that caliph al-Ḥākim was enthroned in 996 at the age of eleven. During the early years of his reign, the caliphate was governed by a vizier. Had Ibn al-Haytham arrived in Cairo before 1000, the construction of the embankments of the Nile would have been ordered by al-Azīz (975–96), al-Ḥasan ibn ʻAmmār (996–7), or Barjawān (997–1000)—not al-Ḥākim himself.

7. Images in Full Distortion

These exclusions leave one possible date, 21 October 990, and several possible locations. To disentangle these cases, I will calculate the distortion of the projected image from the relative azimuth β.

Baṣra 990. The old city of Baṣra—now Zubayr—was established along the watercourse of the Pallakopas, not the Shaṭṭ al-'Arab. The city had several mosques and sūqs, and two libraries: that of Ibn Sawwār and that of 'Aḍud al-Dawla. The latter was destroyed in A.H. 483/1090 (Massignon 1963: 61–87). Although the urban fabric of the city is not perfectly regular, morphological analysis reveals that the façade azimuth ranges from 199 to 263°, with 77 % of buildings facing 232 ± 12°. Therefore, the direction of the city was set by the riverbed, flowing to 139° ($139° - \frac{\pi}{2} = 229°$).

Baghdād$_1$ 990. According to legend, Baghdād's observatory was built next to the House of Wisdom (*Bayt al-Ḥikma*). However, it has since been convincingly established that this link is not based on historical facts. It is now known that the House of Wisdom only operated as: (a) a library, (b) a translation center, and (c) a place of discussion (Sayılı 1960: 53–55; Balty-Guesdon 1992: 132–138). Baghdād's observatory was erected by caliph al-Ma'mūn in ca. 828 near the Shamāsiyya Gate, in a neighborhood whose location is only approximately known because the area fell to ruin during the 11th century (Makdisi 1959: 186). Removing late buildings and basing all calculations on the map of cultivated plots—land ownership is a conservative set of data—provides us with an azimuth ranging from 229° to 284°, with a mean Az_m = 265° and 77 % of the plots facing 265 ± 8°.[44] Because of the uncertainty of the data, the identification of this location must be regarded with caution.

Baghdād$_2$ 990. In a later period, Sharaf al-Dawla ordered al-Qūhī to build an observatory in the garden of the Royal Palace in Baghdād (al-Bīrūnī 1967: 69–70). In A.H. 379/988, several astronomers took part in observations there, including al-Ṣāghānī, Abū al-Wafā' al-Būzjānī, al-Sāmarrī, Abū al-Ḥasān al-Maghribī, and Hilāl al Ṣābī. Since the Royal Palace was erected on the bank of the Tigris, the azimuth $Az = 221°$ is known with confidence.

Cairo 1030. Since it is widely believed that Ibn al-Haytham's *Optics* was a late work, since it is quoted in *On the Shape of the Eclipse*, and since Ibn al-Haytham

---

44. Insofar as many astronomical instruments were oriented along the meridian line, it is likely that several buildings were facing $Az_a = 270°$. The values are close enough, that it is not necessary to add a new case.

came to Egypt under the reign of al-Ḥākim, I add Cairo as a possible place of observation for the partial eclipse of 31 August 1030. I fix the observer's location at the gate of the Azhar mosque-university, in accordance with Ibn al-Qifṭī (1903: 167). Hence, $Az = 309°$.

Whatever the case, I choose the wall so that $\alpha = \widehat{x\mathrm{OP}'}$ is minimum:

$$\alpha_{\text{Bas}} = 232° - \mathrm{Az}_{990} \qquad \alpha_{\text{Bagh}_1} = 265° - \mathrm{Az}_{990}$$

$$\alpha_{\text{Bagh}_2} = 221° - \mathrm{Az}_{990} \qquad \alpha_{\text{Cai}} = 309° - \mathrm{Az}_{1030}$$

The eclipse image can then be calculated (Table App2) and drawn (Fig. App4).

	Baṣra 990	*Baghdād* 1 990	*Baghdād* 2 990	*Cairo* 1030
UT	11:47:54	11:37:16	11:37:16	15:19:14
ΔT[a]	± 56.5	± 56.5	± 56.5	± 50.5
Az_S	234.766	228.699	228.699	271.638
Alt_S	26.423	28.750	28.750	10.659
Az_M	234.832	228.739	228.739	271.752
Alt_M	26.524	28.834	28.834	10.801
ΔAz	+3.930	+2.358	+2.358	+6.840
ΔAlt	+6.060	+5.034	+5.034	+8.478
γ[b]	32.964	25.099	25.099	38.896
$2b_S$	32.331	32.331	32.331	31.872
$2b_M$	29.782	29.800	29.782	31.905
α_S	1.832	36.300	7.699	37.362
β_S	26.524	28.750	28.750	10.659
ϑ_S	26.583	45.043	29.678	38.638
φ_S	63.417	44.957	60.322	51.362
$2a_S$	36.120	45.757	37.212	40.804
λ	+3.665	+47.179	+13.724	+72.768
u_M	+3.932	+2.926	+2.379	+8.606
v_M	+6.767	+5.742	+5.742	+8.627

Table App2. UT in h:mn:sec, ΔT in sec. Other data expressed in decimal degrees, except ΔAz, ΔAlt, $2b$, $2a$, u and v presented in arcmns. All data calculated from Giorgini, NASA Jet Propulsion Laboratory, except [a] Uncertainty ΔT interpolated from tables created by Espenak and Meeus (2006, 13) and [b] Data calculated from Espenak and O'Byrne (2011), NASA Goddard Space Flight Center. Lunar data calculated in the same manner.

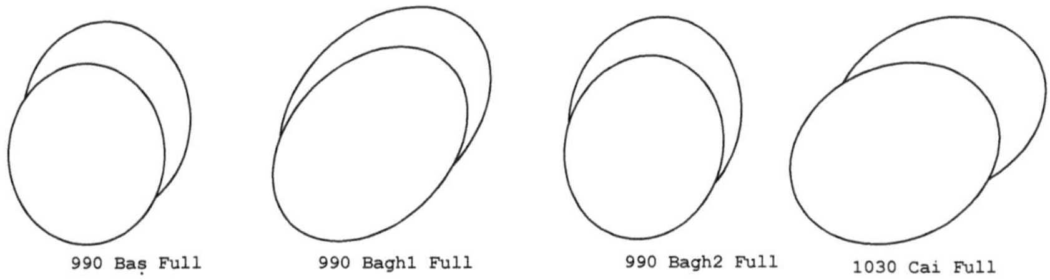

Fig. App4. Eclipse Images in Full Distortion

Whereas Baṣra ($2a_S$ / $2b_S$ = 1.117, λ = +4°) and Baghdād$_2$ (1.141, +14°) are consistent with the diagrams, Baghdād$_1$ (1.415, +47°) and Cairo (1.280, +73°) are not. The latter two locations do not fit because they would create too strong a distortion. In both cases, ellipse orientation and eccentricity conflict with the manuscript diagrams. This provides further proof that Baghdād$_1$ and Cairo must be eliminated as possible sites for the 31 August 1030 observation.

Astronomical analysis alone, however, cannot be used to eliminate Baghdād as a possible location for observing the eclipse. (a) Baghdād$_2$ provides an image with a good fit to those in the manuscripts. (b) Even if observing an eclipse through a pinhole was customary at that time—al-Bīrūnī describes a "light entering through a small hole into a house" (1976, I: 47)—Ibn al-Haytham mentions a special method for observing solar eclipses that was used in Baghdād:

"The eclipse was observed through the apertures made in several parts of the *tharema**; We observed the Sun in water, safely and accurately. We found at the end, when no further part of the Sun was eclipsed, and the disc appeared complete in the water, the altitude to be 12° eastwards."[39]

"If the observer cannot look at the Sun, let him put a bowl where the eclipsed Sun falls, let him pour clear water in it, and expect to be calm. Now, if he looks in [the bowl], he will see the Moon by reflection, and then he will find her [as if it were directly seen] on the face of the Sun."

39. Ibn Yūnus (1804: 114–116, 122). *Tharema is a wooden structure: "tharema, maison de bois: c'est un mot étranger." Ibn al-Haytham's text is provided by Masoumi (2006, I: 99):

فإن لم يستطع الناظر النظر إلى الشمس، فدعه يضع طستًا في موضع من كسف الشمس ويسكب فيه ماءً صافيًا ويصبر إلى أن يسكن الماء، ثم ينظر في الماء فسيرى القمر بالانعكاس ويجده في وجه الشمس.

A historical fact finally dismisses Baghdād as a possible place of observation. The astronomical observations conducted in the garden of the Royal Palace of Baghdād ceased when Sharaf al-Dawla passed away on 7 September 989 (al-Bīrūnī 1967: 70). Shortly after, al-Qūhī moved to Baṣra. His correspondence with Hilāl al-Ṣābī suggests that Sharaf's brother, Bahā' al-Dawla, who succeded him as the ruler of 'Irāq (989–1012), had far less interest in science. In all likelihood, al-Qūhī used Baṣra as a fallback position after the death of his patron Sharaf al-Dawla (Berggren 1983: 72). As Baghdād's observatory was closed, there was no reason for Ibn al-Haytham to have moved from Baṣra to Baghdād at that time. Some scholars speculate further that al-Qūhī and Ibn al-Haytham came into contact with this opportunity (Berggren 1983: 90–91; Hogendijk 1985: 113–115).

8. Discussion

The conclusion that Ibn al-Haytham's *On the Shape of the Eclipse* might have recorded the partial solar eclipse seen on 21 October 990 in Baṣra must be regarded with caution, considering the uncertainty surrounding the use of the two combined methods. This outcome warrants a closer look.

The dating of past astronomical events has been regularly subjected to discussion. Crucial points raised include: (1) Documents used for dating are sometimes too vague to support any valid conclusions. (2) Several cases of forgery have been detected in eclipse records. (3) Contrary to total eclipses (three by millennium), partial solar eclipses occur frequently (385 by millennium) and thus are not easily distinguished. (4) Uncertainty increases with remoteness in time (Courville 1975; Steele 2003; Keenan 2002; a reply in Pankenier 2007). The present dating is free, *a priori*, of criticism based on point 2. It is partly subject to criticism vis-à-vis point 4 due to the geophysical factors that result in uncertainty in the length of day. It is also sensitive to point 3: seven eclipses (970, 990, 1000, 1015, 1030, 1032, 1033) fit the calculated magnitude, two (990, 1030) fit the magnitude and the occultation angle, and one (990) fits the magnitude, occultation angle, and image projection. Finally, as Baghdād$_2$ was partly eliminated on historical grounds, the result is not entirely immune to criticism 1.

Image projection computation can only be attempted if: (5) Essential data, such as the eclipse magnitude and angle of occultation are available, and (6) The place of observation is accurately known. The present dating is open to both criticisms. In terms of point 5, the astronomical data in this study have been collected from hand-drawn diagrams and the dating thus depends on their veracity. Notwithstanding the arguments made in the introduction, research on scientific diagrams is too recent to have formed a definitive view on this issue. Regarding point 6, even if the places chosen for computing the images in full distortion are historically based, they do not represent the only possible locations.

Conclusion

This Appendix has endeavored to date the solar eclipse recorded in *On the Shape of the Eclipse* through a method combining image reconstruction with astronomical data. Building on the assumption that Ibn al-Haytham's diagrams were reality-based, this study applied an ephemeris to sift through the eclipses that occurred during Ibn al-Haytham's lifetime. The remaining cases were sorted by computing the image projection in full or vertical distortion, depending on whether the azimuth was available or not. This procedure shows that *On the Shape of the Eclipse* might have recorded the partial solar eclipse observed on 28 Rajab A.H. 380/21 October 990 CE, most likely from Baṣra.

This finding should be considered as the best possible candidate at present for the date and place of the eclipse recorded by Ibn al-Haytham. This result is tentative pending further research, but if it is confirmed by independent sources, then our approach could offer a useful new tool in the analysis of scientific images that scholars have begun to engage in over the past decades.

The method in itself deserves attention because it has a wide range of potential applications. Even though the results are method- and data-dependent, as in any empirical research, calculating the image projection is efficient for reconstructing what an observer could physically have seen of an eclipse. Let us consider a final example. On December 4, 1639 (November 24, 1639 using the Julian calendar), Jeremiah Horrocks predicted and observed the transit of Venus through a helioscope. He gave a full account of this event in *Venus in sole visa*, published in 1662 (Van Roode 2012). On similar grounds, the image illustrating the book has been proven not to be that of Horrocks, but that of his editor, Johannes Hevelius, because it does not match the image reconstructed from astronomical data.

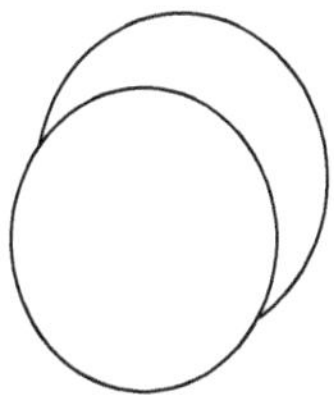

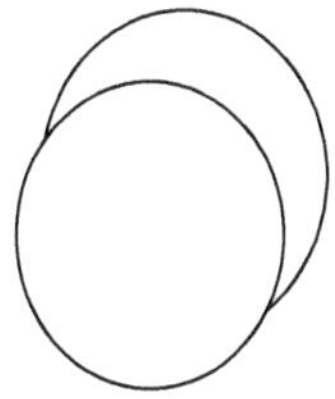

I. BAṢRA

Date	Begins	Alt_B	Max UT	Alt_M	Az	Ends	Alt_E	Magn.	Obsc.
0970 May 08	03:20:34	17	04:23:14	30	086	05:34:19	46	0.622	0.528
0976 Jul 29									
0977 Dec 13	06:53:27	29	08:02:01	35	167	09:14:42	36	0.399	0.286
0978 Jun 08	12:47:52	36	13:48:34	23	284	14:42:47	12	0.334	0.216
0979 May 28									
0985 Jul 20	14:54:27	10	15:22:42	4	291	(s)15:43	(s)0	0.136	0.059
0988 May 18	13:48:36	22	14:33:43	12	287	15:15:29	4	0.274	0.164
0989 Nov 01	12:29:40	16	13:03:24	10	244	13:35:02	4	0.074	0.023
0990 Oct 21	10:06:44	41	11:47:54	26	235	13:13:43	10	0.744	0.657
0992 Mar 07	06:31:33	41	08:09:50	54	159	09:52:03	54	0.703	0.619
0993 Aug 20	06:05:49	47	07:27:13	63	131	08:53:01	70	1.000	1.000
0999 Oct 12	03:42:40	10	04:17:46	17	112	04:54:49	24	0.106	0.040
1000 Apr 07	08:05:49	66	09:30:12	66	206	10:53:54	54	0.891	0.870
1004 Jan 24	13:57:04	5	(s)14:21	(s)0	249	(s)14:21	(s)0	0.388	0.272
1007 May 19	06:18:25	56	07:38:14	72	120	09:03:52	79	0.715	0.652
1013 Jan 14									
1015 Jun 19	02:11:58	4	03:11:53	16	72	04:18:55	30	0.947	0.924
1016 Nov 02									
1017 Oct 22	13:48:50	3	(s)14:03	(s)0	255	(s)14:03	(s)0	0.177	0.085
1020 Aug 21									
1021 Aug 11	(r)02:16	(r)0	02:59:49	9	79	03:51:17	20	0.349	0.234
1023 Jan 24									
1024 Jun 09	(r)01:45	(r)0	(r)01:45	(r)0	62	01:46:39	0	0.010	0.001
1025 May 29									
1030 Aug 31	14:21:20	9	(s)15:03	(s)0	278	(s)15:03	(s)0	0.633	0.543
1032 Jan 15	11:16:55	30	12:29:50	19	231	13:34:43	7	0.813	0.772
1032 Jul 10	06:35:00	58	06:42:16	60	99	06:49:19	61	0.002	0.000
1033 Jun 29	11:41:31	51	13:02:15	34	279	14:11:50	19	0.839	0.793
1037 Apr 18									

Table 3: Eclipse predictions by Fred Espenak and Chris O'Byrne (NASA-GSFC). Hours in Universal Time: Baṣra Local Time = Universal Time+3h 11m 21s. *Alt* (°): Sun's Altitude: (r) Sunrise, (s) Sunset. *Az* (°): Sun's Azimuth. Magn: Eclipse Magnitude (i.e. the fraction of the Sun's diameter occulted by the Moon). Obsc: Eclipse Obscuration (i.e. the portion of the Sun's area obscured by the Moon).

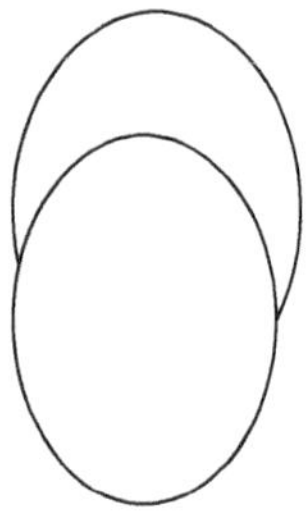

990 Cai Proj

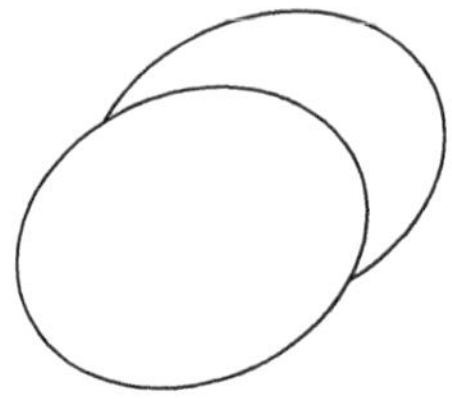

1030 Cai Full

II. CAIRO

Date	Begins	Alt_B	Max UT	Alt_M	Az	Ends	Alt_E	Magn.	Obsc.
0970 May 08	03:19:37	3	04:18:04	15	77	05:23:13	29	0.779	0.717
0976 Jul 29	16:11:52	6	$^{(s)}$16:42	$^{(s)}$0	291	$^{(s)}$16:42	$^{(s)}$0	0.286	0.176
0977 Dec 13	06:20:09	16	07:27:19	26	142	08:42:28	34	0.601	0.511
0978 Jun 08	12:10:53	59	13:35:24	40	276	14:47:51	25	0.501	0.385
0979 May 28	16:09:25	7	$^{(s)}$16:46	$^{(s)}$0	296	$^{(s)}$16:46	$^{(s)}$0	0.411	0.293
0985 Jul 20	14:41:06	26	15:24:50	17	283	16:05:48	8	0.284	0.174
0988 May 18	13:55:57	34	14:28:38	27	279	14:59:29	21	0.100	0.037
0989 Nov 01	12:15:09	30	12:34:34	27	228	12:53:37	24	0.016	0.002
0990 Oct 21	09:25:06	47	11:14:21	42	212	12:53:49	27	0.687	0.591
0992 Mar 07	06:09:07	25	07:29:56	40	125	09:01:17	54	0.541	0.430
0993 Aug 20	05:44:05	29	06:55:56	44	105	08:15:57	60	0.957	0.958
0999 Oct 12									
1000 Apr 07	07:48:10	54	09:01:27	65	147	10:17:53	68	0.575	0.481
1004 Jan 24	13:48:57	19	14:55:44	6	245	$^{(s)}$15:28	$^{(s)}$0	0.981	0.971
1007 May 19	06:02:41	38	07:07:33	52	95	08:19:40	68	0.545	0.448
1013 Jan 14	11:43:03	35	12:16:12	31	215	12:48:10	27	0.079	0.027
1015 Jun 19	$^{(r)}$02:54	$^{(r)}$0	03:14:39	4	65	04:12:57	15	0.786	0.723
1016 Nov 02	14:54:34	1	$^{(s)}$15:01	$^{(s)}$0	251	$^{(s)}$15:01	$^{(s)}$0	0.014	0.002
1017 Oct 22	13:35:28	19	14:53:27	3	253	$^{(s)}$15:09	$^{(s)}$0	0.657	0.556
1020 Aug 21	15:59:41	4	$^{(s)}$16:20	$^{(s)}$0	282	$^{(s)}$16:20	$^{(s)}$0	0.310	0.199
1021 Aug 11	$^{(r)}$03:23	$^{(r)}$0	$^{(r)}$03:23	$^{(r)}$0	74	03:49:09	5	0.308	0.194
1023 Jan 24	12:37:34	30	13:04:47	26	227	13:31:12	22	0.075	0.024
1024 Jun 09									
1025 May 29	16:12:00	7	16:46:19	0	296	$^{(s)}$16:48	$^{(s)}$0	0.249	0.145
1030 Aug 31	14:18:24	24	15:19:14	11	272	$^{(s)}$16:09	$^{(s)}$0	0.660	0.576
1032 Jan 15	10:47:45	39	12:06:26	32	213	13:17:25	22	0.622	0.537
1032 Jul 10									
1033 Jun 29	11:10:11	72	12:41:42	53	269	14:00:00	36	0.645	0.556
1037 Apr 18	08:22:51	63	08:54:15	68	139	09:26:40	72	0.058	0.016

Table 4: Eclipse predictions by Fred Espenak and Chris O'Byrne (NASA-GSFC). Hours in Universal Time: Cairo Local Time = Universal Time+2h 05m 03s. *Alt* (°): Sun's Altitude: $^{(r)}$ Sunrise, $^{(s)}$ Sunset. *Az* (°): Sun's Azimuth. Magn: Eclipse Magnitude (i.e. the fraction of the Sun's diameter occulted by the Moon). Obsc: Eclipse Obscuration (i.e. the portion of the Sun's area obscured by the Moon).

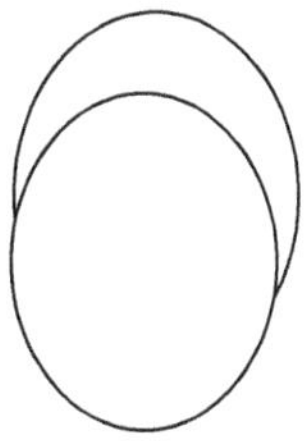

990 Dam Proj

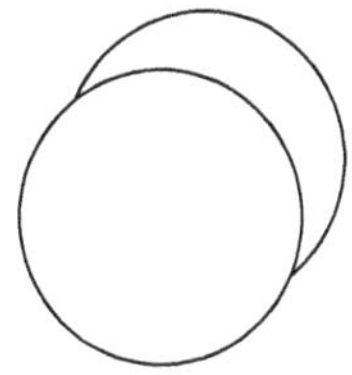

1030 Dam Proj

III. DAMASCUS

Date	Begins	Alt_B	Max UT	Alt_M	Az	Ends	Alt_E	Magn.	Obsc.
0970 May 08	03:22:38	9	04:23:56	21	81	05:32:37	35	0.802	0.746
0976 Jul 29	16:01:15	5	(s)16:28	(s)0	291	(s)16:28	(s)0	0.337	0.233
0977 Dec 13	06:32:44	18	07:37:05	27	150	08:47:38	32	0.461	0.352
0978 Jun 08	12:27:03	51	13:37:16	36	276	14:39:15	23	0.365	0.246
0979 May 28	16:01:00	6	(s)16:35	(s)0	297	(s)16:35	(s)0	0.462	0.346
0985 Jul 20	14:44:53	22	15:19:43	15	284	15:52:55	8	0.181	0.090
0988 May 18	13:39:54	34	14:27:04	24	280	15:10:26	15	0.238	0.134
0989 Nov 01									
0990 Oct 21	09:32:12	44	11:20:40	36	217	12:57:19	21	0.774	0.693
0992 Mar 07	06:23:52	30	07:47:10	44	137	09:18:24	53	0.520	0.407
0993 Aug 20	05:50:41	34	07:04:34	49	114	08:25:25	62	0.973	0.977
0999 Oct 12	(r)03:44	(r)0	04:09:30	5	104	04:42:14	11	0.101	0.037
1000 Apr 07	08:00:36	57	09:15:34	65	169	10:32:17	62	0.605	0.515
1004 Jan 24	13:51:58	13	14:56:09	1	248	(s)15:02	(s)0	0.963	0.948
1007 May 19	06:13:59	45	07:19:48	59	106	08:31:44	72	0.508	0.406
1013 Jan 14	11:48:10	29	12:23:35	25	220	12:57:31	20	0.101	0.038
1015 Jun 19	(r)02:25	(r)0	03:17:20	9	68	04:17:53	21	0.768	0.702
1016 Nov 02									
1017 Oct 22	13:36:19	13	(s)14:45	(s)0	254	(s)14:45	(s)0	0.613	0.506
1020 Aug 21	15:51:24	2	(s)16:04	(s)0	282	(s)16:04	(s)0	0.209	0.112
1021 Aug 11	(r)02:58	(r)0	03:00:00	0	73	03:54:22	11	0.515	0.406
1023 Jan 24	12:43:08	24	13:11:01	20	231	13:38:02	15	0.086	0.030
1024 Jun 09									
1025 May 29	16:16:55	3	(s)16:36	(s)0	298	(s)16:36	(s)0	0.117	0.047
1030 Aug 31	14:13:17	20	15:14:16	7	273	(s)15:51	(s)0	0.748	0.683
1032 Jan 15	11:00:10	34	12:13:54	26	218	13:20:29	16	0.600	0.510
1032 Jul 10									
1033 Jun 29	11:15:09	66	12:44:53	48	268	14:02:04	32	0.791	0.733
1037 Apr 18	08:47:57	67	09:12:37	69	167	09:37:42	69	0.033	0.007

Table 5: Eclipse predictions by Fred Espenak and Chris O'Byrne (NASA-GSFC). Hours in Universal Time: Damas Local Time = Universal Time+2h 25m 14s. *Alt* (°): Sun's Altitude: (r) Sunrise, (s) Sunset. *Az* (°): Sun's Azimuth. Magn: Eclipse Magnitude (i.e. the fraction of the Sun's diameter occulted by the Moon). Obsc: Eclipse Obscuration (i.e. the portion of the Sun's area obscured by the Moon).

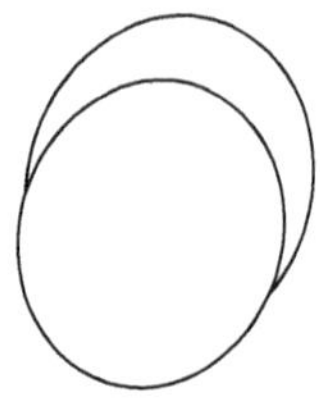

990 Bagh2 Full

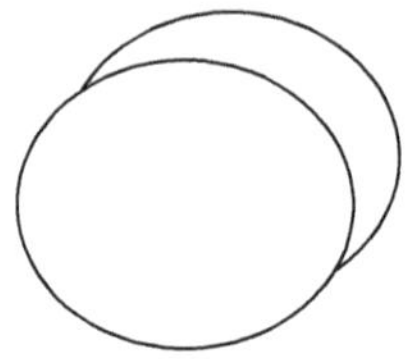

1030 Bagh Full

IV. BAGHDĀD

Date	Begins	Alt_B	Max UT	Alt_M	*Az*	Ends	Alt_E	Magn.	Obsc.
0970 May 08	03:22:45	15	04:26:19	28	086	05:38:10	43	0.718	0.643
0976 Jul 29									
0977 Dec 13	06:49:21	25	07:53:44	31	162	09:02:35	33	0.376	0.263
0978 Jun 08	12:44:09	40	13:43:39	28	281	14:37:06	17	0.297	0.182
0979 May 28	15:57:11	1	(s)16:02	(s)0	297	(s)16:02	(s)	0.081	0.027
0985 Jul 20	14:52:25	14	15:19:25	8	288	15:45:33	3	0.117	0.047
0988 May 18	13:40:41	27	14:29:56	17	284	15:15:12	8	0.310	0.196
0989 Nov 01									
0990 Oct 21	09:52:58	41	11:37:16	29	229	13:07:06	13	0.793	0.714
0992 Mar 07	06:33:36	38	08:05:44	50	153	09:42:48	53	0.606	0.505
0993 Aug 20	06:01:05	43	07:19:09	57	127	08:42:27	67	0.936	0.933
0999 Oct 12	03:34:04	4	04:14:05	12	110	04:56:41	20	0.154	0.069
1000 Apr 07	08:07:39	62	09:28:29	65	195	10:49:19	55	0.760	0.704
1004 Jan 24	13:55:08	6	(s)14:30	(s)0	249	(s)14:30	(s)0	0.559	0.456
1007 May 19	06:20:28	53	07:33:58	68	119	08:53:03	77	0.597	0.509
1013 Jan 14	12:28:27	19	12:33:42	19	229	12:39:11	18	0.003	0.001
1015 Jun 19	02:17:15	4	03:15:55	15	72	04:21:02	28	0.853	0.808
1016 Nov 02									
1017 Oct 22	13:43:02	6	(s)14:13	(s)0	254	(s)14:13	(s)0	0.350	0.230
1020 Aug 21									
1021 Aug 11	(r)02:26	(r)0	03:00:20	7	78	03:55:40	18	0.455	0.341
1023 Jan 24									
1024 Jun 09	(r)01:52	(r)0	(r)01:52	(r)0	61	01:54:37	0	0.030	0.006
1025 May 29									
1030 Aug 31	14:15:47	13	15:13:53	1	277	(s)15:19	(s)0	0.769	0.707
1032 Jan 15	11:12:39	30	12:24:45	20	227	13:29:23	9	0.699	0.630
1032 Jul 10									
1033 Jun 29	11:31:04	56	12:54:57	39	274	14:07:18	24	0.871	0.833
1037 Apr 18									

Table 6: Eclipse predictions by Fred Espenak and Chris O'Byrne (NASA-GSFC). Hours in Universal Time: Baghdād Local Time = Universal Time+2h 57m 33s. *Alt* (°): Sun's Altitude: (r) Sunrise, (s) Sunset. *Az* (°): Sun's Azimuth. Magn: Eclipse Magnitude (i.e. the fraction of the Sun's diameter occulted by the Moon). Obsc: Eclipse Obscuration (i.e. the portion of the Sun's area obscured by the Moon).

References

Abū al-Fidā' al-Ḥamawī (1848) *A Sketch of the Countries* (*Kitāb taqwīm al-buldān*). *Géographie d'Aboulféda*, traduite de l'arabe en français par M. Reinaud. Tome II, première partie, contenant la première moitié de la traduction du texte arabe. Paris: Imprimerie Nationale.

al-Andalusī (1912) *Kitāb ṭabaqāt al-umam* (*Book of the Categories of Nations*), edited by Louis Cheikho. Beirut: Imprimerie Catholique. French transl. *Livre des catégories des nations*, R. Blachère. Paris: Larose, 1935.

Alhacen Latinus, see Smith 2001–2006; Pietquin 2010.

Alhacen Italicus, see Excerpts in Federici Vescovini 1965, 1998.

Arago F. (1839) "Le daguerréotype," *Comptes-rendus des séances de l'Académie des sciences*, séance du lundi 19 avril 1839, tome 9: 250–267.

Aristotle [Pseudo-] (1995) "Problems," *The Complete Works of Aristotle.* The Revised Oxford Translation, Edited by Jonathan Barnes. Princeton: Princeton University Press, II, pp. 1319–1527.

Aristotle [Pseudo-] (1999) *The Problemata Physica Attributed to Aristotle. The Arabic Version of Ḥunain ibn Isḥāq and the Hebrew Version of Moses ibn Tibbon*, ed. L.S. Filius. Leiden/Boston: Brill.

Assemani S.E. (1742) *Bibliothecae Mediceae Laurentianae et Palatinae Codicum Mss. Orientalium Catalogus...*, Florentiae: ex Typographio Albiziano.

Balty-Guesdon M.G. (1992) "Le Bayt al-Ḥikma de Baghdad," *Arabica* 39: 131–150.

Barakat R. (1965) "Rayleigh Wavefront Criterion," *Journal of the Optical Society of America* 55: 572–573.

Barbrook A.C., N. Blake, and P.M.W. Robinson (1998) "The Phylogeny of the Canterbury Tales," *Nature* 394: 839.

al-Bayhaqī (1935), *Tatimmat ṣiwān al-ḥikma* (*Continuation of the Receptacle of Wisdom*), edited by M. Shafī'. Lahore: L. Ishwar Das.

Berggren J.L. (1983) "The Correspondence of Abū Sahl al-Kūhī and Abū Isḥāq al-Sābī. A Translation with Commentaries," *Journal for History of Arabic Science* 7: 39–123.

Bernard C. (1999) *Experimental Medicine.* With an Introduction by S. Wolf. New Brunswick: Transaction Publishers.

D. Raynaud, *A Critical Edition of Ibn al-Haytham's On the Shape of the Eclipse*,
Sources and Studies in the History of Mathematics and Physical Sciences,
DOI 10.1007/978-3-319-47991-0

al-Bīrūnī (1967) *Kitāb taḥdīd nihāyat al-amākin li-taṣḥīḥ masāfat al-masākīn* (*The Determination of Coordinates of Positions for the Correction of Distances Between Cities*), edited by J. 'Alī. Beirut: American University of Beirut.

al-Bīrūnī (1976) *Kitāb ifrād al-maqāl fī amr al-ẓilāl* (*The Exhaustive Treatise on Shadows*), 2 vols. Edited by E.S. Kennedy. Aleppo: Institute for the History of Arabic Science.

el-Bizri N. (2005) "Ibn al-Haytham," T.F. Glick, S. Livesey and F. Wallis, eds., *Medieval Science, Technology and Medicine: An Encyclopedia.* New York/Oxon: Routledge, pp. 237–240.

Björnbo A.A. and S. Vogl (1912) "Alkindi, Tideus und Pseudo-Euclid, drei optische Werke," *Abhandlungen zur Geschichte der mathematischen Wissenschaften* 26: 1–176.

Bonnet-Bidaud J.M., F. Praderie and S. Whitfield (2009) "The Dunhuang Chinese Sky: A Comprehensive Study of the Oldest Known Star Atlas," *Journal of Astronomical History and Heritage* 12: 39–59.

Bosworth C.E. (1975) "Iran under the Buyids," R.N. Frye, ed., *The Cambridge History of Iran, Volume 4; From the Arab Invasion to the Saljuqs.* Cambridge: Cambridge University Press, pp. 250–305.

Bosworth C.E. (1993) "Misāḥa," *Encyclopédie de l'Islam.* Leiden: Brill, VII, pp. 139b–140a.

Bouby L., P. Boissinot and P. Marinval (2011) "Evidence of Beer-brewing in Mediterranean France and the Consumption of Alcoholic Beverages During the 5th Century BC," *Human Ecology* 39(3): 351–360.

Bouguer P. (1760) "Observations faites pour déterminer combien les parties du Soleil voisines du centre, sont plus lumineuses que celles qui sont vers les bords de cet Astre," *Traité d'Optique sur la gradation de la lumière*, ouvrage posthume de M. Bouguer, publié par M. L'Abbé de La Caille. Paris: Guerin & Delatour, pp. 90–6.

Bourgain P. and F. Vielliard (2002) *Conseils pour l'édition des textes médiévaux*, III. Paris: École des Chartes.

Bretagnon P. and G. Francou (1988) "Planetary Theories in Rectangular and Spherical Variables. VSOP87 Solutions," *Astronomy and Astrophysics* 202: 309–315.

Brey G. (2009) "Scientific Manuscripts in the Digital Age." *Digital Proceedings of the Lawrence J. Schoenberg Symposium on Manuscript Studies in the Digital Age. Issue 1: On the Nature of Things: Modern Perspectives on Scientific Manuscripts*, Art. 5: 1–7.

Brosselard C. (1861) "Les inscriptions arabes de Tlemcen, XIV: La coudée royale de Tlemcen," *Revue Africaine* 5: 14–30.

Bussemaker (1857) *Aristotelis Opera Omnia. Graece et Latine cum Indice Nominum et Rerum Absolutissimo. Volumen Quartum: Problematum Sectiones XLI...* Parisiis: Editore Ambrosio Firmin Didot.

Cambiano G. (1992) "La démonstration géométrique." M. Detienne, ed., *Les Savoirs de l'écriture en Grèce ancienne.* Lille: Presses universitaires de Lille, pp. 251–272.

Canard M. (1975) "al-Ḥākim bi Amr Allāh," *Encyclopédie de l'Islam*. Leiden: E.J. Brill, vol. III, pp. 79–84.

Cardelle de Hartmann C., P. Schwagmeier and P. Roelli (2013) *Petrus Alfonsi, Dialogus: Kritische Edition und Kommentar* (work in progress). Zürich: Universität Zürich, Philosophische Fakultät.

Carlsson K. (2004) *The Pinhole Camera Revisited or the Revenge of the Simple-Minded Engineer*, Based on a Seminar Given at KTH (Royal Institue of Technology), Fall 2004. http://www.biox.kth.se/kjellinternet/Pinhole.pdf

Carman C. (2009) Rounding Numbers: Ptolemy's Calculation of the Earth-Sun Distance, *Archive for History of Exact Sciences* 63: 205–242.

Cesariano C. (1521) *Di Lucio Vitruvio de Architectura Libri Decem traducti del Latino in Vulgare affigurati, Commentati, et con mirando ordine Insignati*. Como: Gotardus de Ponte.

Chapront-Touzé M. and J. Chapront (1983) "The Lunar Ephemeris ELP2000," *Astronomy and Astrophysics* 124: 50–62.

Cheikho L. (1908) "Risālat al-Khujandī fī [taṣḥīḥ] al-mayl wa-'arḍ al balad," *al-Mashriq* 11: 60–9.

Cipolla A., M. Buzzoni, O.E. Haugen and R. Rosselli Del Turco, eds. (2012) *IV Incontro di Filologia Digitale: Constitutio Textus* (Verona, 13–15 September 2012).

Clagett M. (1964) *Archimedes in the Middle Ages*, 1. *The Arabo-Latin Tradition*. Madison: The University of Wisconsin Press.

Clagett M. (1978) *Archimedes in the Middle Ages*, 3: *The Fate of the Medieval Archimedes 1300-1565*, II: *The Arabo-Latin and Handbook Traditions of Archimedes in the Fourteenth and Early Fifteenth Centuries*. Philadelphia: American Philosophical Society.

Clagett M. (1980) *Archimedes in the Middle Ages*, 4: *A Supplement on the Medieval Latin Traditions of Conic Sections, 1150-1566*. Philadelphia: The American Philosophical Society.

Clagett M. (1984) *Archimedes in the Middle Ages, 5: Quasi-Archimedean Geometry in the Thirteenth Century*. Philadelphia: The American Philosophical Society.

Courville D.A. (1975) "Limitation of Astronomical Dating Methods," *Kronos* 2: 49–72.

Crozet P. (2005) "Editer les figures des manuscrits arabes de géométrie: l'exemple d'al Sijzī," P. Mascellani, P.D. Napolitani and V. Gavagna, eds., *The Problem of Diagrams and Drawings Criticism in Mathematical Texts*. A Workshop in Pisa, Report 1.0: 33–42.

Danti E. (1573) *La prospettiva di Euclide, nella quale si tratta di quelle cose, che per raggi diritti si veggono & di quelle, che con raggi reflessi nelli Specchi appariscono*. Firenze: Giunti.

Darrigol O. (2012) *A History of Optics from Greek Antiquity to the Nineteenth Century*. Oxford: Oxford University Press.

Decorps-Foulquier M. (1999) "Sur les figures du traité des coniques d'Apollonios de Perge édité par Eutocius d'Ascalon," *Revue d'histoire des mathématiques* 5: 61–82.

Decorps-Foulquier M., Federspiel M., eds. (2008) *Apollonius de Perge, Coniques. Tome 1.2: Livre I*, Berlin: Walter de Gruyter.

Dees A. (1988) "Ecdotique et informatique," J. Kremer, ed., Actes du XVIIIe Congrès international de linguistique et de philologie romanes (Trier, 18–24 May 1986). Tübingen: Niemayer, VI, pp. 18–27.

De Goeje M.J. (1906) *Descriptio Imperii Moslemici* (i.e., Moqaddasī's *The Best Divisions in the Knowledge of the Regions*). Lugduni Batavorum: E.J. Brill.

Delambre J.B.J. (1819) *Histoire de l'astronomie du moyen âge*. Paris: Courcier.

De Young G. (2004) "The Latin Translation of Euclid's Elements Attributed to Gerard of Cremona in Relation to the Arabic Transmission," *Suhayl* 4: 311–383.

De Young G. (2005) "Diagrams in the Arabic Euclidean Tradition," *Historia Mathematica* 32: 129–179.

Donohue J.J. (2003) *The Buwayhid Dynasty in Iraq 334H./945 to 403H./1012. Shaping Institutions for the Future*. Leiden/Boston: E.J. Brill.

Duhem P. (1913) *Le Système du monde. Histoire des doctrines cosmologiques de Platon à Copernic*, vol. 3. Paris: librairie scientifique Hermann.

Eastwood B.S. (1986) "Alhazen, Leonardo, and Late-Medieval Speculation on the Inversion of Images in the Eye," *Annals of Science* 43: 413–446.

Egidius de Baisiu, see Mancha 1989.

Emmel A. (2011) "Pinhole Photography," R. Lambrecht and C. Woodhouse, eds., *Way Beyond Monochrome. Advanced Techniques for Traditional Black & White Photography*. Amsterdam: Elsevier pp. 149–156.

Ettinghausen R. (1942) "Painting in the Fatimid Period: A Reconstruction," *Ars Islamica* 9: 112–124.

Ettinghausen R. (1956) "Early Realism in Islamic Art," *Studi Orientalistici in onore di Giorgio Levi della Vida*. Rome: Istituto per l'Oriente, I, pp. 252–273.

Eustache D. (1955) "Al-Basra, capital idrisside et son port," *Hesperis* 42: 217–238.

Eutocius d'Ascalon (1972) *Archimède, Oeuvres, IV: Commentaires d'Eutocius, Fragments*, texte établi et traduit par Ch. Mugler. Paris: Les Belles Lettres.

Evans J. (1998) *The History and Practice of Ancient Astronomy*. New York: Oxford University Press.

al-Falakī M. [Maḥmūd Bey] (1873) "Le système métrique actuel d'Égypte. Les nilomètres anciens et modernes et les antiques coudées d'Égypte," *Journal Asiatique* 1: 67–110.

al-Farghānī, Abū al-'Abbās Aḥmad ibn Muḥammad ibn Kathīr (1669) *Compendium of Astronomy* (*Kitāb fī jawāmi' 'ilm al-nujūm*). *Muhammedis Fil. Ketiri Ferganensis, qui vulgo Alfraganus dicitur, Elementa Astronomica: arabice & latine*. Opera Jacobi Golii. Amstelodami: apud Johannem Jansonium.

al-Fārisī, Kamāl al-Dīn (1928–9) *Kitāb Tanqīḥ al-Manāẓir li-ḏawī al-abṣār wa al-baṣā'ir*, 2 vols. ed. by M. Ḥijāzī. Hyderabad: Majlis Dā'ira al-Ma'ārif al-'Uthmāniyya.

Federici Vescovini G. (1965) "Contributo per la storia della fortuna di Alhazen in Italia: il volgarizzamento del ms. Vat. lat. 4595 e il 'Commentario terzo' del Ghiberti," *Rinascimento* 5: 17–33.

Federici Vescovini G. (1998) "Alhazen vulgarisé: Le 'De li aspecti' d'un manuscrit du Vatican (moitié du XIVe siècle) et le troisième Commentaire sur l'optique de Lorenzo Ghiberti," *Arabic Science and Philosophy* 8: 67–96.

Felsenstein J. (2009) *Phylip. Phylogeny Inference Package*. Version 3.69. http://evolution.genetics.washington.edu/phylip.html.

Filius L.S. (2006) "The Genre Problemata in Arabic: Its Motions and Changes," P. de Leemans and M. Goyens, eds., *Aristotle's Problemata in Different Times and Tongues*. Leuwen: Leuwen University Press, pp. 33–54.

Galileo (1613) Lettera a Federico Cesi [in Roma], Le Selve, 25 Gennaio 1613. *Le Opere di Galileo Galilei*, edizione nazionale, a cura di A. Favaro. Firenze: Tip. G. Barberà, 1901, vol. XI, pp. 465–9.

Gamalina R.B., Sokolova E.I. (1972) "K istorii geometricheskoj optiki v Blijnem i Srednem Vostoke i v Evrope v srednie veka," S.H. Siraždinov, pod red., *Iz istorii tochnykh nauk na srednevekovom Blijnem i Srednem Vostoke: Sbornik statej*. Akademija Nauk Uzbekskoj SSR. Tachkent: Izdatel'stvo "FAN," pp. 72–75.

Gemma Frisius R. (1545) *De radio astronomico et geometrico liber*. Antwerp: apud Greg. Bontium; 2nd edition Paris: G. Cavellat, 1557.

Glenisson J. ed. (1979) *La Pratique des ordinateurs dans la critique des textes*. Actes du Colloque international (Paris, 29–31 mars 1978). Paris: Éditions du CNRS.

Goldstein B.R. (1985) *The Astronomy of Levi ben Gerson (1288–1344)*. A Critical Edition of Chapters 1–20 with Translation and Commentary. New York/Berlin/Tokyo: Springer.

Goldstein B.R. and G. Hon (2007) "Celestial Charts and Spherical Triangles: The Unifying Power of Symmetry," *Journal for the History of Astronomy* 38: 1–14.

Gorini R. (2003) "Al-Haytham, The Man of Experience. First Steps in the Science of Vision" *Journal of the International Society for the History of Islamic Medicine* 2: 53–55.

Gossellin P.F.J. (1819) *Recherches sur le principe, les bases et l'évaluation des différents systèmes métriques linéaires de l'Antiquité*. Paris: Imprimerie Royale.

Graham A.C. (1978) *Later Mohist Logic, Ethics and Science*. London: University of London School of Oriental and African Studies.

Grasshoff G. (1990) *The History of Ptolemy's Star Catalogues*. New York/Berlin: Springer.

Halma N.B, Delambre J.B.J. (1813) ΚΛΑΥΔΙΟΥ ΠΤΟΛΕΜΑΙΟΥ ΜΑΘΗΜΑΤΙΚΗ ΣΥΝΤΑΞΙΣ. *Composition mathématique de Claude Ptolémée* traduite du grec en français par M. Halma, et suivie de notes de M. Delambre, Paris: H. Grand.

Hamilton R.W. (1946) "An Eighth-Century Water-Gauge at al-Muwaqqar," *Quarterly of the Departement of Antiquities in Palestine* 12: 70–2.

Heath T.L. (1912) *The 'Method' of Archimedes Recently Discovered by Heiberg. A Supplement to the Works of Archimedes 1897*. Cambridge: Cambridge University Press.

Hestroffer D., Magnan C. (1998) "Wavelength Dependency of the Solar Limb Darkening," *Astronomical Journal* 333: 338–42.

Hinz W. (1955) *Islamische Masse und Gewichte: umgerechnet ins metrische System*. Leiden, Köln: E.J. Brill.

Hinz W. (1965) "Dhirāʿ," *Encyclopédie de l'Islam*. Leiden: E.J. Brill, vol. II, pp. 238b–239a.

Hogendijk J.P. (1984) "Greek and Arabic Constructions of the Regular Heptagon," *Archive for History of Exact Sciences* 30: 197–330.

Hogendijk J.P. (1985) *Ibn al-Haytham's Completion of the Conics*. New York/Berlin/Tokyo: Springer.

Hogendijk J.P. (1986) "Discovery of an 11th Century Geometrical Compilation: the Kitab al-Istikmāl of Yūsuf al-Mu'taman ibn Hūd, King of Saragossa," *Historia Mathematica* 13: 43–52.

Hogendijk J.P. and Sabra A.I., eds. (2003) *The Enterprise of Science in Islam*. Cambridge, MA: The MIT Press.

Huygens R.B.C. (2001) *Ars edendi. Introduction pratique à l'édition des textes latins du Moyen Age*. Turnhout: Brepols.

Ibn Abī Uṣaybiʿa (1882–4), *'Uyūn al-anbā' fī ṭabaqāt al-aṭibbā'* (*Sources of Information on the Generations of Physicians*), 2 vols., ed. A. Müller. Cairo: al-Maṭbaʿa al-Wahbiyya.

Ibn al-Nadīm (1970) *The Fihrist of Ibn al-Nadim. A Tenth-Century Survey of Muslim Culture*. English Translation by B. Dodge. New York: Columbia University Press.

Ibn al-Qifṭī (1903) *Ta'rīkh al-ḥukamā'* (*The History of Learned Men*), edited by J. Lippert. Leipzig: Dieterich Verlag.

Ibn ʿAsākir (2000) *Tā'rīkh Dimashq* (*History of Damascus*). Beirut: Dār al-Fikr.

Ibn Yūnus (1804) *Zīj al-kabīr al-Ḥākimī* (*The Book of the Great Hakemite Table*). French Translation by Caussin, *Le Livre de la Grande Table Hakémite*. Paris: Imprimerie de la République.

Janssens J.L. (2004) "Experience (*tajriba*) in Classical Arabic Philosophy," *Quaestio* 4: 45–62.

Jardine B., N. Jardine (2010) "Critical Editing of Early Modern Astronomical Diagrams," *Journal for History of Astronomy* 41(3): 393–414.

Jaubert P.A. (1836–40) *Géographie d'Édrisi traduite de l'arabe en français d'après deux manuscrits de la Bibliothèque du Roi et accompagnée de notes*. Paris: Imprimerie nationale.

Keenan D.J. (2002) "Astro-Historiographic Chronologies of Early China Are Unfounded," *East Asian History* 32: 61–68.

Kennedy E.S. (1991-2) "Transcription of Arabic Letters in Geometric Figures," *Zeitschrift für Geschichte der Arabisch-Islamischen Wissenschaften* 7: 21–22.

Kepler I. (1600) *Eclipse Notebook* [St. Petersburg: Archive of the Russian Academy of Sciences, Pulkovo Collection, MS XV, fol. 247 sq.] *Johannes Kepler Gesammelte Werke*, Bd II, hrsg. von Franz Hammer. München: C.H. Beck'sche, 1939.

Kepler I. (1604) *Ad Vitellionem paralipomena, quibus Astronomiae pars optica traditur...* Frankfurt: C. Marnius.

Khalidov A.B. (1986) *Arabskie rukopisi Instituta vostokovedenija kratkij katalog, chast' 2: Ukazateli i priloženie.* Moscow: Izdatel'stvo "Nauka."

Kheirandish E. (1996) "The Arabic 'Version' of Euclidean Optics: Transformations as Linguistic Problems in Transmission," F.J. Ragep and S.P. Ragep and S. Livesey, eds., *Tradition, Transmission, Transformation: Proceedings of Two Conferences on Pre-modern Science Held at the University of Oklahoma.* Leiden: Brill, 227–243.

Kheirandish E. (1999) *The Arabic Version of Euclid's Optics: Kitāb Uqlīdis fī Ikhtilāf al-Manāẓir*, Edited and Translated wih Historical Introduction and Commentary, 2 vols. New York: Springer-Verlag.

Kheirandish E. (2003) "The Many Aspects of 'Appearances': Arabic Optics to 950 AD," Hogendijk J.P. and Sabra A.I., eds., *The Enterprise of Science in Islam.* Cambridge, MA: The MIT Press, 55–84.

Kheirandish E. (2009) "Footprints of 'Experiment' in Early Arabic Optics," *Early Science and Medicine* 14(1): 79–104.

al-Khujandī, see Cheikho 1908; Schirmer 1926–7; Wiedemann 1910.

Kircher A. (1646) *Ars magna lucis et umbrae in decem libros digesta.* Rome: apud H. Scheus.

Knorr W.R. (1989) *Textual Studies in Ancient and Medieval Geometry.* Boston-Basel-Berlin: Birkhäuser.

Kraemer J.L. (1992) *Humanism in the Renaissance of Islam: The Cultural Revival During the Buyid Age.* Leiden/Boston: Brill.

Krause M. (1936) "Stambuler Handschriften islamischer Mathematiker," *Quellen und Studien zur Geschichte der Mathematik, Astronomie und Physik, Studien* 3: 437–532.

Langermann Y.T. (1984) "The Mathematical Writings of Maimonides," *Jewish Quartely Review* 75: 57–65.

Langermann Y.T. (1996) "Medieval Hebrew Texts on the Quadrature of the Lune," *Historia Mathematica* 23: 31–53.

Langermann Y.T. (2000) "Some Remarks on Judah ben Solomon ha-Cohen and his Encyclopedia, Midrash ha-Ḥokhmah," S. Harvey, ed., *The Medieval Hebrew Encyclopedias of Science and Philosophy.* Dordrecht: Kluwer, 371–389.

Lefèvre W. ed. (2003) *Inside the Camera Obscura. Optics an Art under the Spell of the Projected Image.* Berlin: Max-Planck-Institute für Wissenschaftsgeschichte, Preprint 333.

Levi ben Gerson, see Goldstein 1985.

Lévy T. (1992) "Gersonide, commentateur d'Euclide: Traduction annotée de ses gloses sur les Éléments," Freudenthal G., ed., *Studies on Gersonides: A Fourteenth-Century Jewish Philosopher-Scientist*. Leiden: E.J. Brill, 87–88.

Lévy T. (1997) "The Establishment of the Mathematical Bookshelf of the Medieval Hebrew Scholar: Translations and Translators," *Science in Context* 10(3): 431–451.

Libri G. (1841) *Histoire des sciences mathématiques en Italie, depuis la Renaissance des Lettres jusqu'à la fin du dix-septième siècle*, tome 4. Paris: Renouard.

Lindberg D.C. (1968) "The Theory of Pinhole Images from Antiquity to the Thirteenth Century," *Archive for History of Exact Sciences* 5(2): 154–176.

Lindberg D.C. (1970*a*) "A Reconsideration of Roger Bacon's Theory of Pinhole Images," *Archive for History of Exact Sciences* 6(3): 214–223.

Lindberg D.C. (1970*b*) "The Theory of Pinhole Images in the Fourteenth Century," *Archive for History of Exact Sciences* 6(4): 299–325.

Lindberg D.C. (1996) *Roger Bacon and the Origins of Perspectiva in the Middle Ages*. A Critical Edition and English Translation, with Introduction and Notes. Oxford: Clarendon Press; New York: Oxford University Press.

Lindberg D.C. (1998) *Roger Bacon's Philosophy of Nature*. A Critical Edition, with English Translation, Introduction, and Notes, of the 'De multiplication specierum' and 'De speculis comburentibus.' South Bend: St. Augustine's Press.

Lommel E. von (1885) "Die Beugungserscheinungen einer kreisrunden Öffnung und eines kreisrunden Schirmchens theoretisch und experimentell bearbeitet," *Abhandlungen der Mathematisch-Physikalischen Classe der Königlich Bayerischen Akademie der Wissenschaften* 15: 229–328.

Lommel E. von (1886) "Die Beugungserscheinungen geradlinig begrenzter Schirme," *Abhandlungen der Mathematisch-Physikalischen Classe der Bayerischen Akademie der Wissenschaften* 15: 529–664.

Long J. (1979), *Bartholomaeus Anglicus, De proprietatibus rerum, Books 3–4: On the Properties of Soul and Body*. Toronto: Centre for Medieval Studies by the Pontifical Institute of Medieval Studies.

Loth O. (1877) *A Catalogue of the Arabic Manuscripts in the Library of the India Office*. London, no editor.

Maas P. (1958) *Textual Criticism*. Oxford: Clarendon Press.

Maas P. (2010) "Computer Aided Stemmatics. The Case of Fifty-Two Text Versions of Carakasa saṃhitā Vimānasthāna 8.67–157," *Wiener Zeitschrift für die Kunde Südasiens* 52/53: 63–120.

Macé C., T. Schmidt and J.F. Weiler (2001) "Le classement des manuscrits par la statistique et la phylogénétique," *Revue d'Histoire des Textes* 31: 241–273.

Macé C. and P.V. Baret (2006) "Why Phylogenetic Methods Work: The Theory of Evolution and Textual Criticism," C. Macé et al., eds., *The Evolution of Texts. Confronting Stemmatological and Genetical Methods*. Roma-Pisa: Istituti editoriali e poligrafici internazionali, pp. 89–108.

Makdisi G. (1959) "The Topography of Eleventh Century Baġdād," *Arabica* 6: 178–197.

Mancha J.L. (1989) "Egidius of Baisiu's theory of pinhole images," *Archive for History of Exact Sciences* 40: 1–35.

Manders K. (2008) "The Euclidean Diagram," P. Mancosu, ed., *The Philosophy of Mathematical Practice*. Oxford: Oxford University Press, pp. 80–133.

Mascellani P., P.D. Napolitani and V. Gavagna, eds. (2005) *The Problem of Diagrams and Drawings Criticism in Mathematical Texts*. A Workshop Held in Pisa, Report 1.0.

Massignon L. (1963), "Explication du plan de Baṣra, Irak," *Opera Minora*, vol. III. Edited by Youakim Moubarac. Beirut: Dar al-Maaref.

Masoumi Hamedani H. (2006) *L'Optique et la physique céleste. L'oeuvre optico-cosmologique d'Ibn al-Haytham*, 2 vols. Doctoral Dissertation. Paris: Université Paris 7.

Maçoudi (1861) *Les Prairies d'or* (*Murūj al-dhahab wa ma'ādin al-jawhar*), vol. 1. Texte et traduction par Barbier de Meynard et Pavet de Courteille. Paris: Imprimerie impériale.

Matvievskaja G.P., Rozenfeld B.A. (1983) "Al-Hasan ibn al-Hajsam," *Matematiki i astronomy musul'manskogo srednevekov'ja i ikh trudy (VIII-XVII vv.)* 2. *Matematiki i astronomy, vremja žizni kotorykh isvestno*. Moscow: Izdatel'stvo "Nauka," pp. 240–255.

Mercier R.P. (1984) "The Astronomical Tables of Rajah Jai Singh Sawā'i," *Indian Journal of History of Science* 19: 143–171.

Mercier R.P. (1994) "Geodesy," J.B. Harley and D. Woodward, eds., *The History of Cartography, Vol. 2, Book 1: Cartography in the Traditional Islamic and South Asian Societies*. Chicago and London: The University of Chicago Press, 175–88.

Mersenne (1651) *L'Optique et la Catoptrique du Reverend Pere Mersenne Minime*, nouvellement mise en lumiere apres la mort de l'Autheur. Paris: Langlois.

Meyerhof M. (1948) "'Alī al-Bayhaqī's Tatimmat Ṣiwān al-Ḥikma. A Biographical Work on Learned Men of the Islam," *Osiris* 8: 122–217.

Mielenz K.D. (1999) "On the Diffraction Limit of Lensless Imaging," *Journal of Research of the National Institute of Standards and Technology* 104: 479–485.

Mooney L.R., A.C. Barbrook, C.J. Howe and M. Spencer (2001) "Stemmatic analysis of Lydgate's Kings of England: A test case for the application of software developed for evolutionary biology to manuscript stemmatics," *Revue d'Histoire des Textes* 31: 275–297.

Mumma J., M. Panza and G. Sandu, eds. (2013) "Diagrams in Mathematics: History and Philosophy," *Synthese* 186, 1: 7–20.

Naẓīf M. (1942–3) *Al-Ḥasan Ibn al-Haytham wa-buḥūthuhu wa-kushūfuhu al-baṣariyyah* (*Al-Ḥasan Ibn al-Haytham: His Research and Discoveries in Optics*), 2 vols. al-Qāhirah:

Nūrī Press. Reprint with an Introduction by Roshdi Rashed. Bayrūt: Markaz dirāsāt al-waḥdah al-'arabiyyah, 2008.

Netz R. (1999) *The Shaping of Deduction in Greek Mathematics: A Study in Cognitive History*. Cambridge: Cambridge University Press.

Needham J. et al. (1962) *Science and Civilisation in China, 4: Physics and Physical Technology, Part I: Physics*. Cambridge: Cambridge University Press.

Nyman G., J. Radun, T. Leisti et al. (2006) "What Do Users Really Perceive: Probing the Subjective Image Quality", L.C. Cui and Y. Miyake, eds., *Image Quality and System Performance, III* (January 15, 2006). San Jose, CA: SPIE 6059.

Omar S.B. (1977) *Ibn al-Haytham's Optics: A Study of the Origins of Experimental Science*. Minneapolis: Bibliotheca Islamica.

Orléans, Chérubin d' (1671) *Dioptrique oculaire ou la Théorique, la positive, et la méchanique de l'oculaire dioptrique en toutes ses espèces*. Paris: Jolly et Bernard.

Oudet J.F. (1994) "Le principe de la chambre noire et les sextants monumentaux de Rayy (Xe s.) et de Samarkand (XVe s.)," *Comprendre et maîtriser la nature au Moyen Âge. Mélanges d'histoire des sciences offerts à Guy Beaujouan*. Genève: Droz, pp. 29–53.

Pankenier D.W. (2007) "Caveat Lector: Comments on Douglas J. Keenan," *Journal of Astronomical History and Heritage* 10: 137–141.

Pedersen F.S. (2014) *William of Saint-Cloud: Almanach Planetarum. An Edition of the Canons, A Few Samples from the Tables and a Foray into the Numbers*. Copenhagen: Centre for the Aristotelian Tradition.

Petro L.D., Foukal P.V., Rosen W.A. et al. (1984) "A Study of Solar Photospheric Limb-Darkening Variations," *Astrophysical Journal* 283: 426–38.

Porta, G.B. della (1558) *Magiae Naturalis libri viginti ab ipso quidem authore adaucti, nunc vero ab infinitis, quibus editio illa scatebat mendis, optime repurgati; in quibus scientiarum Naturalium divitiae et deliciae demonstrantur...* Neapoli: H. Salvianum.

Pérez J.P. (2000) *Optique. Fondements et applications*. Paris: Dunod.

Petzval J. (1857) "Camera-Obscura-Objective," *Sitzungsberichte der Kaiserlichen Akademie der Wissenschaften, Mathematisch-Naturwissenschaftlichen Classe* 26: 33–90.

Pietquin P. (2010) *Le Septième livre du traité De aspectibus d'Alhazen, traduction latine médiévale de l'Optique d'Ibn al-Haytham*. Bruxelles: Académie royale de Belgique.

Popper W. (1951) *The Cairo Nilometer. Studies in Ibn Taghrī Birdī's Chronicle of Egypt*. Berkeley and Los Angeles: University of California Press.

Popper W. (1955) *Egypt and Syria under the Circassian Sultans, 1382–1468 A.D. Systematic Notes to Ibn Taghrī Birdī's Chronicles of Egypt*, 2. Berkeley and Los Angeles: University of California Press.

Prell H. (1960) "Die schwarzen Ellen der Araber," *Zeitschrift der Deutschen Morgenländischen Gesellschaft* 110: 26–42.

al-Qūshjī, 'Alī (1652) *Tract on Astronomy (Risāla dar 'ilm al-hay'a)*. "Ali Kushgius De Terrae magnitudine, & sphaerarum coelestium a terra distantiis," *Astronomica quaedam ex traditione Shah Cholgii Persae* [ed. J. Greaves]. Londini: J. Flesher, pp. 93–97.

Rashed R. (1993) *Les Mathématiques infinitésimales du IXe au XIe siècle, vol. II: Ibn al-Haytham*. London: al-Furqān Foundation.

Rashed R. (1997) *Oeuvres philosophiques et scientifiques d'al-Kindī, vol. 1: L'Optique et la catoptrique*. Leiden/New York: E.J. Brill.

Rashed R. (2000) *Les Mathématiques infinitésimales du IXe au XIe siècle, vol. III: Ibn al-Haytham. Théorie des coniques, constructions géométriques et géométrie pratique*. London: al-Furqān Foundation.

Rashed R. (2002) *Les Mathématiques infinitésimales du IXe au XIe siècle, vol. IV: Ibn al-Haytham. Méthodes géométriques, transformations ponctuelles et philosophie des mathématiques*. London: al-Furqān Foundation.

Rashed R. (2006) *Les Mathématiques infinitésimales du IXe au XIe siècle, vol. V: Astronomie, géométrie sphérique et trigonométrie*. London: al-Furqān Foundation.

Rashed R. (2005) *Geometry and Dioptrics in Classical Islam*. London: al-Furqān Islamic Heritage Foundation.

Rashed R. (2007) "The Configuration of the Universe: a Book by al-Hasan Ibn al-Haytham?" *Revue d'Histoire des Sciences* 60: 47–63.

Rayleigh, Lord [Strutt, J.W.] (1879) "Investigations in Optics, with Special Reference to the Spectroscope," *Philosophical Magazine* 8: 261–274, 403–411, 477–486.

Rayleigh, Lord [Strutt, J.W.] (1891) "Some Applications of Photography", *Nature* 44(1133): 249–254.

Raynaud D. (2003) "Ibn al-Haytham sur la vision binoculaire: un précurseur de l'optique physiologique," *Arabic Sciences and Philosophy* 13: 79–99.

Raynaud D. (2007) "Le tracé continu des sections coniques à la Renaissance, applications optico-perspectives, héritage de la tradition mathématique arabe," *Arabic Sciences and Philosophy* 17: 299–345.

Raynaud D. (2009) "Why Did Geometrical Optics Not Lead to Perspective in Medieval Islam?" M. Cherkaoui and P. Hamilton, eds., *Raymond Boudon. A Life in Sociology*. Oxford: Bardwell Press, 1, pp. 243–266.

Raynaud D. (2012*a*) "Abū al-Wafā' Latinus? A Study of Method," *Historia Mathematica* 39: 34–83.

Raynaud D. (2012*b*) "A Hitherto Unknown Treatise on Shadows Referred to by Leonardo da Vinci, S. Dupré and J. Peiffer, eds., *Perspective as Practice. An International Conference on the Circulation of Optical Knowledge*, Max Planck Institut für Wissenschaftsgeschichte, Berlin, October 12-13, 2012. In press.

Raynaud D. (2013*a*) "Leonardo, Optics, and Ophthalmology," F. Fiorani and A. Nova, eds., *Leonardo da Vinci and Optics*. Venezia: Marsilio Editore, pp. 255–276.

Raynaud D. (2013*b*) "Determining the Speed of Light (1676–1983): An Internalist Study in the Sociology of Science," *Cairn International/L'Année sociologique* 63: 359–398.

Raynaud D. (2014*a*) *Optics and the Rise of Perspective. A Study in Network Knowledge Diffusion.* Oxford: Bardwell Press.

Raynaud D. (2014*b*) "A Tentative Astronomical Dating of Ibn al-Haytham's Solar Eclipse Record," *Nuncius* 29: 324–358.

Raynaud D. (2014*c*) "Building the Stemma Codicum from Geometrical Diagrams. A Treatise on Optics by Ibn al-Haytham as a Test Case," *Archive for History of Exact Sciences* 68: 207–239.

Raynaud D. (2015) "Un fragment du De speculis comburentibus de Regiomontanus copié par Toscanelli et inséré dans les carnets de Leonardo (Codex Atlanticus, 611rb/ 915ra)," *Annals of Science* 72-3: 306–36.

Renner E. (1995) *Pinhole Photography. Rediscovering a Historic Technique.* Boston and London: Focal Press.

Robinson P.M.W. (1996) "Computer-Assisted Analysis and 'Best-Text' Historical Editing," Van Reenen et al., eds., *Studies in Stemmatology.* Amsterdam: Benjamins, pp. 123–134.

Robinson P.M.W. and R. J. O'Hara (1996) "Cladistic Analysis of an Old Norse Manuscript Tradition," *Research in Humanities Computing* 4: 115–137.

Robinson P.M.W. (1998) *New Methods of Editing, Exploring, and Reading The Canterbury Tales.* A talk given at the conference 'I nuovi orizzonti della filologia'. Accademia Nazionale dei Lincei (Rome, May 28, 1998).

Roos T. and T. Heikkilä (2009) "Evaluating Methods for Computer-Assisted Stemmatology Using Artificial Benchmark Data Sets," *Literary & Linguistic Computing* 24(4): 417–433.

Ronchi V. (1970) *The Nature of Light: A Historical Survey,* translated by V. Barocas. London: Heinemann (1st edition: *Storia della luce.* Bologna: Zanichelli, 1939).

Rozenfeld B.A. (1975) "A Medieval Physico-Mathematical Manuscript Newly Discovered in the Kuibyshev Regional Library," *Historia Mathematica* 2: 67–69.

Rozenfeld B.A. (1976) "The List of Physico-Mathematical Works of Ibn al-Haytham Written by Himself," *Historia Mathematica* 3: 75–76.

Rozenfeld B.A. (1984) "Astronomija stran Islama," *Istoriko-Astronomicheskie Issledovanija* 17: 67–122 [Ibn al-Hajsam: 81–82].

Rüchardt E. (1958) *Light, Visible and Invisible.* Ann Arbor: University of Michigan Press.

Sabra A.I. (1971) "The Astronomical Origin of Ibn al-Haytham's Concept of Experiment," *Actes du XIIe Congrès international d'histoire des sciences*, Paris: Blanchard, vol. 3A, pp. 133–136.

Sabra A.I. (1972) "Ibn al-Haytham," C.C Gillispie, ed., *Dictionary of Scientific Biography.* New York: Scribner's Sons, VI, pp. 189–210.

Sabra A.I. (1977) "A Note on Codex Biblioteca Medicea-Laurenziana Or. 152," *Journal for the History of Arabic Science* 1: 276–283.

Sabra A.I. (1978) ""Sensation and Inference in Alhazen's Theory of Visual Perception," Machamer P. and Turnbull R.G., eds., *Studies in Perception: Interrelations in the History of Philosophy and Science*, Columbus: Ohio State University Press, pp. 169–185.

Sabra, A.I. (1983) *Kitāb al-Manāẓir: al-maqālāt 1–2–3 fī al-ibṣār 'alā al-istiqāmah li-al-Ḥasan ibn Haytham* (*The Optics of Ibn al-Haytham, Books I–II–III: On Direct Vision*). al-Kuwayt: al-majlis al-waṭanī lil-thaqāfah wa-al-funūn wa-al-adāb (The National Council for Culture, Arts and Letters).

Sabra A.I. (1989) *The Optics of Ibn al-Haytham, Books I–II–III: On Direct Vision*, 2 vols. London: The Warburg Institute.

Sabra, A.I. (1995–6) "On Seeing the Stars, II," *Zeitschrift für Geschichte der arabisch-islamischen Wissenschaften* 10: 1–59.

Sabra A.I. (1998) "One Ibn al-Haytham or Two? An Exercise in Reading the Bio-Bibliographical Sources" *Zeitschrift für Geschichte der Arabisch-Islamischen Wissenschaften* 12: 1–50.

Sabra A.I. (2002) *Kitāb al-Manāẓir: al-maqālāt al-rābi'ah wa al-khāmisah fī in'ikās al-aḍwā' wa mawāḍi' al-khayālāt al-mubṣara bi-al-in'ikās* (*The Optics of Ibn al-Haytham, Books IV–V: On Reflection and Images Seen by Reflection*). al-Kuwayt: al-majlis al-waṭanī lil-thaqāfah wa-al-funūn wa-al-adāb (The National Council for Culture, Arts and Letters).

Sabra A.I. (2002–3) "One Ibn al-Haytham or Two? Conclusion" *Zeitschrift für Geschichte der Arabisch-Islamischen Wissenschaften* 15: 95–108.

Sabra A.I. (2003) "Ibn al-Haytham's Revolutionary Project in Optics: The Achievement and the Obstacle," Hogendijk J.P. and Sabra A.I., eds., *The Enterprise of Science in Islam*. Cambridge, MA: The MIT Press, pp. 85–118.

Sabra A.I. (2007) "Alhazen's Optics in Europe: Some Notes on What it Said and What it Did Not Say," W. Lefèvre, ed., *Inside the Camera Obscura*. Berlin: Max-Planck-Institute für Wissenschaftsgeschichte, Preprint 333, pp. 53–58.

Sabra A.I. and A. Heinen (1991–2) "On Seeing the Stars I," *Zeitschrift für Geschichte der arabisch-islamischen Wissenschaften* 7: 31–72.

Sabra A.I. and N. Shehaby (1971) *Ibn al-Haytham, al-Shukūk 'alā Baṭlamyūs (Dubitationes in Ptolemaeum)*. Cairo: The National Library Press.

Saint-Cloud, Guillaume of, see Pedersen 2014.

Saigey J.F. (1834) *Traité de métrologie ancienne et moderne, suivi d'un précis de chronologie et des signes numériques*. Paris: Hachette.

Saito K. (2005) "The Diagrams in Codex P of Euclid's Elements," P. Mascellani, P.D. Napolitani and V. Gavagna, eds., *The Problem of Diagrams and Drawings Criticism in Mathematical Texts*, A Workshop Held in Pisa, Report Version 1.0, pp. 19–28.

Saito K. (2006) "A Preliminary Study in the Critical Assessment of Diagrams in Greek Mathematical Works," SCIAMVS 7: 81–144.

Saito K. ed. (2011) *Diagrams in Greek Mathematical Texts.* KAKENHI Grants-in-Aid for Scientific Research, Japan Society for the Promotion of Science. Report Version 2.03.

Saito K. ed. (2013) *Reproduced Diagrams from Greek and Arabic Manuscripts.* Research Report 'Databasing the Manuscript Diagrams of Sources in Ancient and Medieval Mathematics.' Japan Society for the Promotion of Science Grants-in-Aid, No. 2130325.

Saito K. and N. Sidoli (2012) "Diagrams and Arguments in Ancient Greek Mathematics: Lessons Drawn from Comparisons of the Manuscript Diagrams With Those in Modern Critical Editions," K. Chemla, ed., *The History of Mathematical Proof in Ancient Traditions.* Cambridge: Cambridge University Press, pp. 135–162.

Salemans B.J.P. (1996) "Cladistics or the resurrection of the method of Lachmann. On building the stemma of Yvain," P. van Reenen et al., eds., *Studies in Stemmatology.* Amsterdam: J. Benjamins, pp. 3–70.

Salemans B.J.P. (2000) *Building Stemmas with the Computer in a Cladistic, Neo-Lachmannian Way. The Case of Fourteen Text Versions of Lanseloet van Denemerken.* Thesis. Nijmegen: Katholieke Universiteit.

Santbech D. (1561) [Additions to] J. Regiomontanus, *De triangulis planis et sphaerius libri quinque.* Basileae: H. Petri et P. Pernam.

Sauvaire H. (1886) "Matériaux pour servir à l'histoire de la numismatique et de la métrologie musulmanes. Quatrième partie: mesures de longueur," *Journal Asiatique* 8: 479–536.

Sayanagi K. (1967) "Pinhole Imagery," *Journal of the Optical Society of America* 57–9: 1091–1098.

Sayılı A. (1960) *The Observatory in Islam and its Place in the General History of the Observatory.* Ankara: Türk Tarih Kurumu Basımevi.

Scheiner C. (1630) *Rosa Ursina sive Sol ex admirando Facularum et Macularum suarum Phœnomeno varius.* Bracciano: Andrea Phæum.

Schirmer O. (1926–7) "Studien zur Astronomie der Araber," *Sitzungsberichte der physikalisch medizinische Sozietät in Erlangen* 58–59: 33–88 (transl. 63–79).

Schott G. (1657) *Magia Universalis Naturae et Artis, sive Recondita naturalium & artificialium rerum scientia...* Herbipoli: H. Pigrin.

Schramm M. (1963) *Ibn al-Haythams Weg zur Physik.* Wiesbaden: Franz Steiner.

Schwarzschild K. (1906) "Ueber das Gleichgewicht der Sonnenatmosphäre," *Nachrichten von der Gesellschaft der Wissenschaften zu Göttingen, Math.-Phys. Klasse* 41–53.

Schwenter D. (1636) *Deliciae Physico-Mathematicae oder mathematische und philosophische Erquickstunden.* Nuremberg: Endter.

Sédillot L.A. (1844) *Mémoire sur les instruments astronomiques des Arabes.* Paris: Imprimerie Royale.

Sezgin F. and E. Neubauer, eds. (2011) *Science and Technology in Islam*, III.6: *Catalogue of the Collection of Instruments of the Institute for the History of Arabic-Islamic Sciences.* Franckfurt: Institut für Geschichte der Arabisch-Islamischen Wissenschaften.

Sharpe L.T., A. Stockman, W. Jagla and H. Jägle (2005) "A Luminous Efficiency Function V*(λ), for Daylight Adaptation," *Journal of Vision* 5: 947–968.

Sidoli N. (2007) "What We Can Learn from a Diagram: The Case of Aristarchus's On the Sizes and Distances of the Sun and Moon," *Annals of Science* 64: 527–547.

Sidoli N. and J.L. Berggren (2007) "The Arabic version of Ptolemy's Planisphere or Flattening the Surface of the Sphere: Text, Translation, Commentary," SCIAMVS 8: 37–139.

Sidoli N. and K. Saito (2009) "The Role of Geometrical Construction in Theodosius's Spherics," *Archive for History of Exact Sciences* 63: 581–609.

Sidoli N. and C. Li (2011) *The Manuscript Diagrams of al-Harawī's Version of Spherics.* Research report for 'Databasing the Manuscript Diagrams of Sources in Ancient and Medieval Mathematics.' Japan Society for the Promotion of Science Grants-in-Aid, 2009-2010, No. 2130325.

Simon J.L., Chapront-Touzé M., Morando B. and W. Thuillot, eds. (2012) *Introduction aux éphémérides astronomiques. Supplément explicatif à la connaissance des temps.* Les Ulis: EDP Sciences.

Smith A.M. (2001) *Alhacen's Theory of Visual Perception.* A Critical Edition, with English Translation and Commentary, of the First Three Books of Alhacen's De aspectibus, the Latin Version of Ibn al-Haytham's Kitāb al-Manāẓir, 2 vols. Philadelphia: American Philosophical Society.

Smith A.M. (2005) "Reflection on the Hockney-Falco Thesis," *Early Science and Medicine* 10: 163–186.

Smith A.M. (2006) *Alhacen on the Principles of Reflection.* A Critical Edition, with English Translation and Commentary, of Books 4 and 5 of Alhacen's *De Aspectibus*, the Medieval Latin version of Ibn-al-Haytham's *Kitāb al-Manāẓir*, 2 vols. Philadelphia: American Philosophical Society.

Steele J.M. (2000) "A Re-Analysis of the Eclipse Observations in Ptolemy's Almagest," *Centaurus* 42: 89–108.

Steele J.M. (2003) "The Use and Abuse of Astronomy in Establishing Ancient Chronologies," *Physics in Canada* 59: 243–248.

Steinschneider M. (1893) "Miscellen," *Monatsschrift für die Geschichte und Wissenschaft des Judentums* 38: 520–522.

Steinschneider M. (1956) *Die Hebräischen Übersetzungen des Mittelalters und die Juden als Dolmetscher.* 2nd Edition: Graz: Akademische Druck- u. Verlagsanstalt.

Stephenson F.R. and S.S. Said (1991) "Precision of Medieval Islamic Eclipse Measurements," *Journal for the History of Astronomy* 22: 195–207.

Stephenson F.R. and J.M. Steele (2006) "Astronomical Dating of Babylonian Texts Describing the Total Solar Eclipse of S.E. 175," *Journal for the History of Astronomy* 37: 55–69.

Straker S. (1970) Kepler's Optics: A Study in the Foundations of 17th-Century Natural Philosophy. An Unpublished Dissertation. Bloomington: Indiana University.

Straker S. (1981) "Kepler, Tycho, and the 'Optical Part of Astronomy'. The Genesis of Kepler Theory of Pinhole Images," *Archive for History of Exact Sciences* 24(4): 267–293.

Suzuki T. (2011) "The Diagrams of the Phaenomena in Greek and Arabic Manuscripts," K. Saito, ed., Diagrams in Greek Mathematical Texts. KAKENHI Grants-in-Aid for Scientific Research, Japan Society for the Promotion of Science. Report version 2.03, pp. 15–38.

Swerd F.M. (1835) *Die Beugungserscheinungen aus den Fundamentalgesetzen der Undulationstheorie. Analytisch entwickelt und in Bildern dargestellt.* Mannheim: Schwan und Goetz.

Swofford D.L. (2003) *PAUP* Phylogenetic Analysis Using Parsimony*, Version 4.0. Tallahassee: Florida State University.

Toomer G.J. (1984) *Ptolemy's Almagest*, translated and annotated by G.J. Toomer. New York/Berlin/Heidelberg/Tokyo: Springer Verlag.

Thro E.B. (1994) "Leonardo da Vinci's Solution to the Problem of the Pinhole Camera," *Archive for History of Exact Sciences* 48: 343–371.

Van Helden A. (1985) *Measuring the Universe: Cosmic Dimensions from Aristarchus to Halley.* Chicago: The University of Chicago Press.

Van Reenen P., M. van Mulken, J. Dyk, eds. (1996) *Studies in Stemmatology.* Amsterdam: John Benjamins.

Van Reenen P., A. den Hollander and M. van Mulken, eds. (2004) *Studies in Stemmatology II.* Amsterdam: John Benjamins.

Van Roode S., "What Did Horrocks Really See?" http://transitofvenus.nl/wp/2012/01/25

Vernet J. (1971) "Ibn al-Haytham," *Encyclopédie de l'Islam.* Leiden: Brill, III: pp. 811–812.

Very F.W. (1902) "The Absorptive Power of the Solar Atmosphere," *Astrophysical Journal* 16: 73–91.

Viré G. (1986) *Informatique et classement des manuscrits. Essai méthodologique sur le 'De astronomia' d'Hygin.* Bruxelles: Éditions de l'Université de Bruxelles.

Wiedemann E. (1906) "Ibn al Haiṭam, ein arabischer Gelehrter," *Festschrift J. Rosenthal zur Vollendung seines siebzigsten Lebensjahres gewidmet.* Lepizig: Thieme, pp. 149–78.

Wiedemann E. (1910) "Zu Ibn al Haiṭams Optik," A*rchiv fur die Geschichte der Naturwissenschaften und der Technik* 3: 1–53.

Wiedemann E. (1914) "Über der Camera obscura bei Ibn al Haiṭam," *Sitzungsberichte phys.-med. Sozietät in Erlangen* 46: 155–169.

Wiedemann E. (1936) "Kamāl al-Dīn al-Fārisī," M.T. Houtsma et al., eds., *E.J. Brill's First Encyclopaedia of Islam*. Leiden: E.J. Brill, 4, p. 704.

Windram H.F., P. Shaw, P. Robinson and C.J. Howe (2008) "Dante's Monarchia as a Test Case for the Use of Phylogenetic Methods in Stemmatic Analysis," *Literary and Linguistic Computing* 23: 443–463.

Winter H.J.J. (1954) "The Optical Researches of Ibn Al-Haitham," *Centaurus* 3: 190–210.

Winter H.J.J. and 'Arafāt W. (1949) "Ibn al-Haitham on the Paraboloidal Focusing Mirror," *Journal of the Royal Asiatic Society of Bengal* 15: 25–40.

Winter H.J.J. and 'Arafāt W. (1950) "A Discourse on the Concave Spherical Mirror by Ibn al-Haitham," *Journal of the Royal Asiatic Society of Bengal* 16: 1–16.

Woerther F. and H. Khonsari (2001) "L'application des programmes de reconstructions phylogénétique sur ordinateur à l'étude de la tradition manuscrite d'un texte: l'exemple du Pseudo-Denys d'Halicarnasse," *Revue d'Histoire des Textes* 31: 227–240.

Würschmidt J. (1914) "Zur Theorie der Camera Obscura bei Ibn al-Haiṯam," *Sitzungsberichte der Physikalisch-medizinischen Sozietät in Erlangen* 46: 151–154. Reprint in Fuat Sezgin et al., *Optics: Texts and Studies*, 2. Frankfurt am Main: Institute for the History of Arabic-Islamic Science, 2001, pp. 253–257.

Young M. (1971) "Pinhole Optics," *Applied Optics* 10: 2763–2767.

Young M. (1972) "Pinhole Imagery," *American Journal of Physics* 40: 715–720.

Young M. (1989) "The Pinhole Camera," *The Physics Teacher* Dec 1989: 648–655.

Youschkevitch A.P. (1961) *Istorija matematiki v srednie veka*. Akademija Nauk SSSR, Institut istorii estestvoznanija i tekhniki, Gosudarstvennoe Izdatel'stvo Fiziko-matematicheskoj Literatury. Moscow: "Fizmatgiz." German translation: [Juschkewitsch A.P.], *Geschichte der Mathematik im Mittelalter*. Leipzig: Teubner, 1964; French partial translation: *Les mathématiques arabes*, VIIIe–XVe siècles. Paris: Vrin, 1976.

Youschkevitch A.P. (1967) "Traditions archimédiennes en mathématiques au Moyen Âge," *Organon* 4: 95–103.

Youschkevitch A.P. (1969) *Fiziko-matematicheskie nauki v stranakh Vostoka, Sbornik statej i publikatsij* 2(5), A.T. Grigor'jan and A.P. Youshkevitch, eds., Akademija Nauk SSSR, Institut istorii estestvoznanija i tekhniki. Moscow: "Nauka."

Zahn J. (1685) *Oculus artificialis teledioptricus, sive Telescopium, ex abditis rerum naturalium & artificialium principiis protractrum...* Würzburg: Quirinus Heyl.

Index Nominum

D. Raynaud, *A Critical Edition of Ibn al-Haytham's On the Shape of the Eclipse*,
Sources and Studies in the History of Mathematics and Physical Sciences,
DOI 10.1007/978-3-319-47991-0

Index Rerum

D. Raynaud, *A Critical Edition of Ibn al-Haytham's On the Shape of the Eclipse*,
Sources and Studies in the History of Mathematics and Physical Sciences,
DOI 10.1007/978-3-319-47991-0

Units of Length: Conversion Chart[1]

1 *dhirā' al-malik* (royal cubit) or *al-hāshimiyya*[2] = 1⅓ *dhirā' al-shar'iyya* (legal cubit)[3] = 2⅔ *shibr* (span) = 8 *qabḍa* (palm) = 32 *iṣba'a* (digit) = 192 *sha'īra* (barleycorn) = 1152 *sha'ar al-khīl* (horsehair)

1 *dhirāʿ al-sharʿiyya* (legal cubit)[4] or *al-yad* or *al-yūsufiyya* = 2 *shibr* (span) = 6 *qabḍa* (palm) = 24 *iṣbaʿa* (digit) = 144 *shaʿīra* (barleycorn) = 864 *shaʿar al-khīl* (horsehair)
1 *shibr* (span) = 3 *qabḍa* (palm) = 12 *iṣbaʿa* (digit) = 72 *shaʿīra* (barleycorn) = 432 *shaʿar al-khīl* (horsehair)
1 *qabḍa* (palm) = 4 *iṣbaʿa* (digit) = 24 *shaʿīra* (barleycorn) = 144 *shaʿar al-khīl* (horsehair)
1 *iṣbaʿa* (digit)[5] = 6 *shaʿīra* (barleycorn) = 36 *shaʿar al-khīl* (horsehair)
1 *shaʿīra* (barleycorn) = 6 *shaʿar al-khīl* (horsehair)

[1] After al-Farghānī (ca. 857), al-Masʿūdī (ca. 940), al-Muqaddasī (ca. 985), *Kitāb al-ḥāwī* (XIth), al-Idrīsī (1154), Abū al-Fidā' (ante 1321) and ʿAlī al-Qūshjī (1457)
[2] Abū'l-Fidā': *al-dhirāʿ al-qadamā'* "the cubit of the Ancients"
[3] Except *Kitāb al-ḥāwī:* 1 *dhirāʿ al-malik* = $1+\frac{1}{8}+\frac{1}{10}$ i.e. $\frac{98}{80}$ black cubit
[4] Abū'l-Fidā': *al-dhirāʿ al-ḥadīthīn* "the cubit of the Moderns," Ibn Khordādbeh provides 1 *dhirāʿ* = 27 digits instead
[5] Except al-Masʿūdī: 1 *iṣbaʿa* = 7 $\frac{2}{9}$ i.e. $\frac{65}{9}$ barleycorns

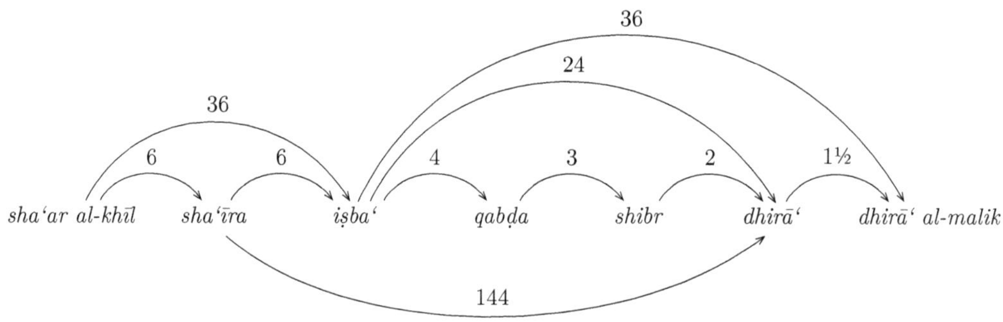

Units of Length: Metric Estimates in Order of Decreasing Size[1]

- Cubit rod of Amenemope, Turin (ante 1190 B.C.): 1 *mḥ nsw* (royal cubit) of 523.4 mm = $\frac{6}{7}$ *mḥ nḏs* (small cubit) of 448.6 mm = 6 *šsp* (palm) of 74.8 mm = 4 *ḏb3* (digit) of 18.69 mm, with divisions from $\frac{1}{2}$ up to $\frac{1}{16}$ [6 grains of 3.11 mm] (inv. 6347).
- Cubit rod of Maya, Paris (1336–27 B.C.): 1 *mḥ nsw* (royal cubit) of 523.0 mm = 7 *šsp* (palm) of 74.7 mm = 4 *ḏb3* (digit) of 18.60 mm, with further divisions from $\frac{1}{2}$ up to $\frac{1}{16}$ [6 grains of 3.10 mm] (inv. N1538).
- Cubit rod of Nippur, Istanbul (2650 B.C.): 1 *kuš* (cubit) of 518.4 mm = 2 *zipaḫ (*span) of 259.2 mm = 15 *šu-si* (digit) of 17.28 mm = 6 *še* (barleycorn) of 2.88 mm (Rottländer 1998).
- Cubit rod of the statue of Gudea, Paris (2120 B.C.): 1 *kuš* (cubit) of 493.3 mm = 2 *zipaḫ* (span) of 246.6 mm = 15 *šu-si* (digit) of 16.44 mm = 6 *še* (barleycorn) of 2.74 mm (Mercier 1994).
- Experimental measurement of 50 samples of six grains of organic barley: 1 *dhirāʿ al-malik* (royal cubit) of 532.2 mm = 1 $\frac{1}{3}$ *dhirāʿ al-sharʿiyya* (legal cubit) of 399.1 mm = 6 *qabḍa* (palm) of 66.5 mm = 4 *iṣbaʿa* (digit) = 16.63 ± 0.81 mm = 6 *shaʿīra* (barleycorn) of 2.77 ± 0.13 mm.
- Royal cubit of the market, Tlemcen (A.D. 1328): 1 *dhirāʿ al-malik* (royal? cubit) of 470.0 mm = 1 $\frac{1}{2}$ *dhirāʿ al-sharʿiyya* (legal cubit) of 313.3 mm = 2 *shibr* (span) of 156.7 mm = 3 *qabḍa* (palm) of 52.2 mm = 4 *iṣbaʿa* (digit) of 13.06 mm = 6 *shaʿīra* (barleycorn) of 2.18 mm = 6 *shaʿar al-khīl* (horsehair) of 0.363 mm (Brosselard 1861).

- Standard of the nilometer on al-Rawḍa (A.D. 641–1522): 1 *dhirā‘ al-malik* (royal cubit) of 616.0 mm = 1 ⅓ *dhirā‘ al-shar‘iyya* (legal cubit) of 462.0 mm = 2 *shibr* (span) of 231.0 mm = 3 *qabḍa* (palm) of 77.0 mm = 4 *iṣba‘a* (digit) of 19.25 mm = 6 *sha‘īra* (barleycorn) of 3.21 mm = 6 *sha‘ar al-khīl* (horsehair) of 0.535 mm (Popper 1951).
- Standard of the water gauge at al-Muwaqqar (A.D. 723): 1 *dhirā‘ al-malik* (royal cubit) of 600.0 mm = 1 ⅓ *dhirā‘ al-shar‘iyya* (legal cubit) of 450.0 mm = 2 *shibr* (span) of 225.0 mm = 3 *qabḍa* (palm) of 75.0 mm = 4 *iṣba‘a* (digit) of 18.75 mm = 6 *sha‘īra* (barleycorn) of 3.13 mm = 6 *sha‘ar al-khīl* (horsehair) of 0.521 mm (Hamilton 1946).
- As the mean *sha‘īra* ranges from 2.82 ± 0.47 mm (Islam) to 2.89 ± 0.34 mm (all values), past values were, in all likelihood, below the current estimates.

[1] The starting point of each series is underlined. While I was studying the literature (Gossellin 1819; Sauvaire 1886; al-Falakī 1873; Hamilton 1946; Popper 1951–55; Hinz 1955; Prell 1960 and Mercier 1994), I was surprised to see more than one author seeking number coincidences—which all provide inconclusive circular arguments—instead of looking for archaeological evidence.

Arabic-English Glossary

ا

one	أحد
to take, to pick	أخذ
to compose, to assemble	ألف
composed (ratio)	مؤلف
Earth	أرض
first, beginning	أول

ب

to demonstrate, to prove, to show	برهن
demonstration, proof	برهان
with proof	بالبرهان
to show by demonstration	تبين بالبرهان
with proof and experimentation	بالبرهان والاعتبار
that is what we wanted to demonstrate	وذلك ما أردنا أن نبين
to see	بصر
vision, eyesight	بصر
Ptolemy	بطلميوس
to disappear	بطل
to be away, to move away, recede	بَعُدَ
distance	بُعْد
opening of the compass	بُعْد
remoteness	بُعْد

D. Raynaud, *A Critical Edition of Ibn al-Haytham's On the Shape of the Eclipse*,
Sources and Studies in the History of Mathematics and Physical Sciences,
DOI 10.1007/978-3-319-47991-0

after, later on	بَعْدَ
far, distant, remote	بَعيد
farther, further	أبعَد
some, part, certain, several	بعض
to maintain, to remain	بقى
what remains of it	ما تبقى منها
to reveal, to appear	بان (بين)
to demonstrate, to explain, to show	بيّن
clear, apparent	بيّن
to show, to discern, to be clear	تبيّن
to show by demonstration	تبيّن بالبرهان

ت

below	تحت
to recite	تلا
to follow	أتلي
next, following	تالي
to be completed, to be finished, to be carried out	تمّ
to complete, to carry out	تمّم

ث

hole, aperture	ثقب
narrow aperture	ثقب ضيق
diameter of the aperture	قطر الثقب
eighth	ثمن
eight	ثمانية
eighteen times	ثمانية عشر مرة
eighteen times and four fifths	ثمانية عشر مرة وأربعة أخماس مرة
eighteen parts and four fifths	ثمانية عشر جُزءًا وأربعة أخماس

ج

it must, to have to, to necessitate	وجب
to cut, to shear	جزّ
part, piece	جزء

body	جسم
self-luminous body	جسم مضيء
to put, to position, to arrange, to set up	جعل
to collect	جمع
all, whole, entire	جميع
to gather	جمل
totality, whole, total	جملة

ح

to track	حاش
track, margin, edge	حواشي
to be curved, domed	حدب
convex	محدّب
convexity	محدبة
to move, to rotate	حرك
movement, motion, variation	حركة
sense	حسّ
to the sense	للحس، في الحسة، عند الحس
perceptible i.e., sensitive	محسوس
imperceptible i.e., insensitives	غير محسوس
to obtain, to get, to receive	حصّل
obtained	حصّل
to surround, to bound, to enclose	حيط
surrouding, bounding	محيط
circumference	محيط
surround	إحاطة

خ

to come out	خرج
to shape	خرط
cone	مخروط

opposite cones	مخروطات متقابلة
Book of Conics	كتاب المخروطات
to draw	خطّ
line, straight line	خط
imaginary lines	خطوط متخيلة
straight line	خط مستقيم
to be hidden	خفي
hidden, invisible	خفيّ
to disappear, to vanish	يختفي
to dissent	خلف
different	اختلاف
contrary, conflicting	مخالف
contrary in position	مخالف لوضع
difference	اختلاف
perceptible difference	اختلاف محسوس
to be hollow	خمص
hollow, dug i.e., deep	انخماص
	د
to insert	دخل
inside, within	داخل
to overlap	تداخل
crossed, overlapping	متداخل
to reach, to get, to perceive	أدرك
perceived by the sense	الذي يدرك الحس
to rotate, to turn	دور
circle	دائرة
circularity, roundness, rotation	استدارة
circular	مستدير
	ذ
to mention, to recall	ذكر
to go	ذهب
going	ذاهب

ر

apex, tip, summit	رأس
quarter	ربع
four	أربعة
forty	أربعون
square	مربع
to fit or press together, to be compact	رصّ
compact, dense, tight	متراصّ
to compound, to aggregate	ركب
compound, aggregate	مركب
to plant, to peg	ركز
center	مركز

ز

to increase	زاد
increase	زيادة
increase and decrease	الزيادة والنقصان
to deviate, to shift	زوي
angle	زاوية
acute angle	زاوية حادة
right angle	زاوية قائمة

س

to hide, to conceal	ستر
hidden, concealed	مستتر
surface, plane, area	سطح
enveloppe surface of the cone	سطح المخروط
conic surface	سطح مخروطي
plane, plane surface	سطح مستو
arrow	سهم
axis of a cone	سهم
generatrix of a cone	سهم
to be equal	سوي
equal	مساوي

ش

strong	شديد
stronger, more intense	أشد
people	شعب
to scatter i.e., to ramify, to fork	يتشعَب
barley	شعير
month	شهر
in the early or last days of the months	في أوائل الشهور وأواخرها
to spread	شعّ
ray	شعاع
to be like	شكل
figure, form, proposition	شكل
Sun	شمس
diameter of the Sun	قطر الشمس
distance of the Sun	بُعْد الشمس
Sun sphere	كرة الشمس

ص

to be true	صحّ
truth, exactitude	صحّ
sound, healthy	صحي
healthier	أصحّ
perfect, true, exact	صحيح
smallness	صغر
small	صغير
less, smaller	أصغر
image, shape, form	صورة
image of the shape	صورة الشكل
image of the crescent	صورة الهلال
to turn into, to become	صير

ض

to weaken	ضعف
weakness	ضعف
weak, dim	ضعيف

to decay, to disappear, to fade away	اضْمحلّ
to shine, to illuminate	ضاء
light	ضوء
sunlight, light of the Sun	ضوء الشمس
sunbeams [in plural]	أضواء الشمس
dim light	ضوء ضعيف
accidental lights	أضواء عرضية
moonlight, light of the Moon	ضوء القمر
lightening	المضيء
luminous	مضيء
self-luminous body	جسم مضيء
self-luminous crescent	هلال مضيء
light crescent	هلال مضيء
narrow	ضيق
narrowness	ضيقه

ط

end, extremity, tip	طرف
tip of the crescent	طرف الهلال
endpoint of the axis	طرف السهم

ظ

shadow	ظل
to rise, to appear	ظهر
visible	ظاهر

ع

to experiment, to test	اعتبر
experimented, tested	اعتبرت
experimentation, experiment	اعتبار
experimenter	المعتبر
to happen	حدث
to happen	عرض
to show, to display	عرض
shown, displayed	عرض
accidental, occasional	عرضي
accidental lights	أضواء عرضية

ten	عشرة
ten times	عشرة أضعاف
tenth	عشر
to increase	عظم
magnitude, greatness	عظم
great	عظيم
greater, greatest	أعظم
cause	علة
to know	علم
science	علم
mathematicians	أصحاب التعاليم
to support	عمد
column, pillar, vertical line	عمود
perpendicular, vertical	عمودي
to be interested	عني
that is to say, I mean	أعني
notion, meaning	معنى

غ

end, goal	غاية
to sink	غرق
to guzzle	استغرق
if it is not a total eclipse	إذا لم يستغرق الكسُوف جميعها
to cover	غطي
to change	غير
to change, to alter	تغير
change, alteration	تغيير

ف

to suppose, to assess	فرض
supposition, hypothesis	فرض
by hypothesis	بالفرض
supposed, imposed	مفروض

to separate, to cut	فصل
separation, section, part, chapter	فصل
intersection i.e., common separation	فصل مشترك
separated	منفصل
to fail	فات (فوت)
difference, disparity	تفاوت
different, uneven	متفاوت
above	فوق

ق

to accept, to host	قبل
to face, to be opposite	قابل
facing, opposite	مقابل
opposite cones	مخروطات متقابلة
to calculate, to assess	قدر
size, measure	قدر
size, measure, magnitude	مقدار
significant, suitable	مقتدر
to precede, to introduce	قدم
lemma, introduction	مقدمة
to move close	قرب
closeness, proximity	قرب
close, near	قرب
closer	أقرب
close	قريبة
approximation	تقريب
to divide	قسم
to divide in halves	قسم بنصفين
to be divided, split	انقسم
to intend, to turn his mind	قصد
country, territory	قطر
diameter, diagonal	قطر
half-diameter, radius	نصف قطر

to cut, to intersect, to describe قطع
    cut, intersected, described قطع
piece, section, sector قطعة

to sit, to take place قعد
base قاعدة

hollow قعر
concavity تقعر
concave مقعر

to diminish, to decrease قلّ
less, fewer أقل

Moon قمر
    diameter of the Moon قطر القمر
    distance of the Moon بُعْد القمر

to arch قوس
arc قوس

to say قال
discourse, epistle, book مقالة

to erect, to raise قام
raised قائم
perpendicular قائم
right angle قائمة
straightness استقامة
    in a straight line, rectilinearly على استقامة

ك

to thicken كثف
dense, thick كثيف
sphere كرة
    sphere of the Sun كرة الشمس
    sphere of the Moon كرة القمر

to darken كسف
eclipse كسوف

total eclipse	كسوف جميع
if this is not a total eclipse	إذا لم يستغرق الكسوف جميعَها
eclipsed	انكسفت
to detect, to reveal, to uncover	كشف
detected, revealed	انكشفت

ل

to stick to, to follow	لزم
necessary	يلزم
to annihilate, to eliminate	لاشى
to vanish, to fade out	تلاشى
night	ليل
clear nights	ليالي يسيرة

م

equal	مثل
such as	مثل
the Almagest	المجسطي
to lengthen	مدّ
to continue, to last	تمادى
to be prolonged	امتدّ
prolonged	ممتد
to pass, to go through	مرّ
passed through	مرت
to touch	مسّ
to be tangent	ماسّ
tangency	تماس
tangent	مماسّ
circle tangent	دائرة التي تماس
to fill	ملأ
filled	امتلأت
to distinguish	ميز
one hundred	مائة (مئة)

ن

to go down	نزل
position, status	منزلة
to resemble, to be like	بمنزلة
to link, to relate	نسب
ratio	نسبة
proportional	كنسبة
half	نصف
to divide into halves	قسم بنصفين
midpoint	منتصف
to look at, to investigate	بحث
to study, to look at	نظر
perspective	مناظر
optics	علم المناظر
Book of Optics	كتاب في المناظر
to pass through, to extend	نفذ
mark, drop	نقط
point	نقطة
midpoint	منتصف
to reach	نهي
end, extremity	نهاية
to end on, to hit [by projection]	نتهي
to stop [a circular motion]	نتهي
ended, hit	انتهى
endlessly, infinitely	متناهية

ه

to begin i.e., the lunation	هلّ
crescent, crescent-shape figure, sickle	هلال
self-luminous crescent	هلال مضيء
light crescent	هلال مضيء
image of the crescent	صورة الهلال

the first [lower] crescent	الهلال الأول
the middle crescent	الهلال الأوسط
the last [upper] crescent	الهلال الأخير
to fall	هوى
air	هواء

و

to require, to be necessary	وجب
to find, to exist	وجد
found	وجدت
to face	وجه
side, direction	جهة
on the side of	في جهة
facing	مواجهة
to be parallel	وزي
parallel	مواز
middle	وسط
to bisect [to be in the middle of]	في وسط
the bisector [the line which is in the middle of]	الخط الذي في وسط
to widen, to expand	وسع
capacity, magnitude, fullness	سعة
wide, large	واسع
to describe, to characterize	وصف
property, character, nature, kind	صفة
to join, to connect	وصل
to join, to connect, to carry on	واصل
chord [the line joining the tips of the crescent]	خط واصل بين طرفي
adjacent, contiguous; persistent, uninterrupted, continuous	متّصل
interlocked with each other	متّصل بعضها ببعض
to place	وضع
position, place, situation	موضع

to adapt	وفق
to agree, to be suited	يوافق
what was agreed	ما اتفق
time, moment	وقت
at the time of the eclipse	في وقت كسوف
imagination, illusion	وهم
to imagine, to hypothesize	توهم

ي

to be easy	يسر
small, easy	يسير

Table of Figures

Figures

Chapter 1

Chapter 2

Chapter 3

D. Raynaud, *A Critical Edition of Ibn al-Haytham's On the Shape of the Eclipse*,
Sources and Studies in the History of Mathematics and Physical Sciences,
DOI 10.1007/978-3-319-47991-0

Chapter 4

Appendix

Plates

Chapter 3

Chapter 4

Appendix

Tables

Credits

All diagrams, photographs and plates are by the author unless otherwise specified.

Plates

D. Raynaud, *A Critical Edition of Ibn al-Haytham's On the Shape of the Eclipse*,
Sources and Studies in the History of Mathematics and Physical Sciences,
DOI 10.1007/978-3-319-47991-0

1. Alhacen Latinus (12th c.)

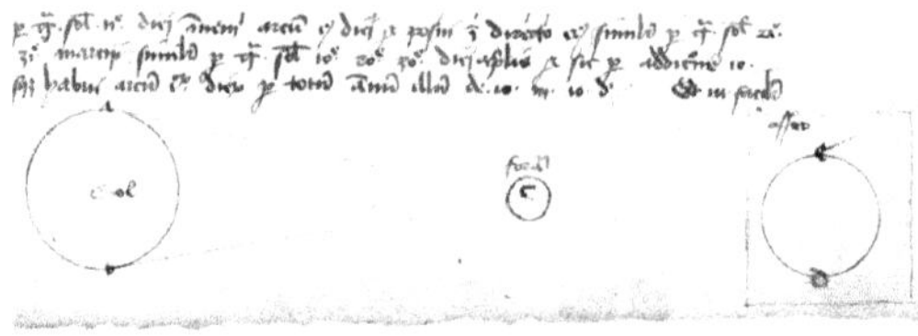

3. Saint-Cloud (1292)

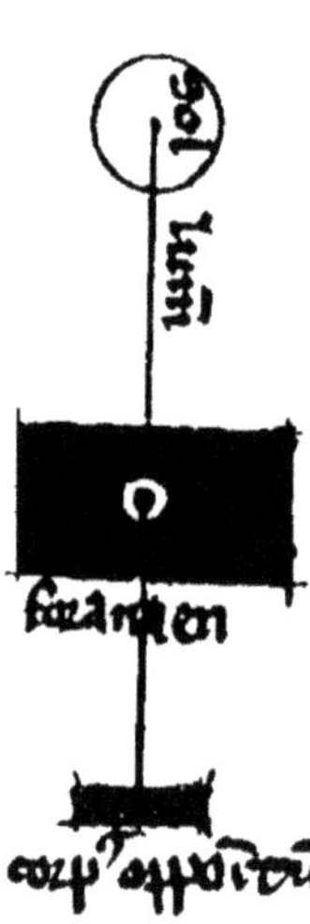

2. Alhacen Latinus (12th c.)

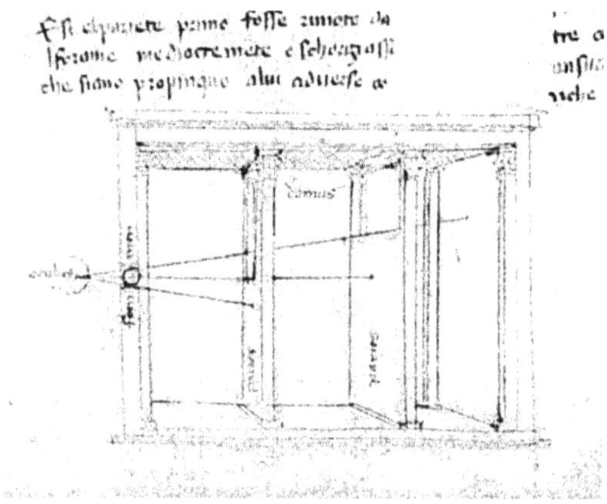

4. Alhacen Italicus (14th c.)

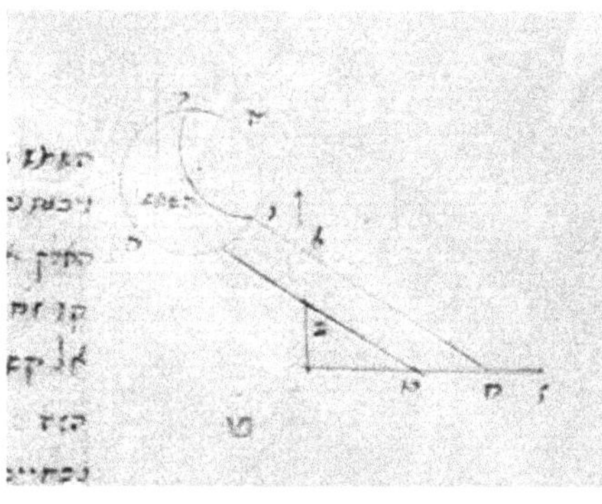

5. Ps.-Euclid, Mantova ebr.3 (?)

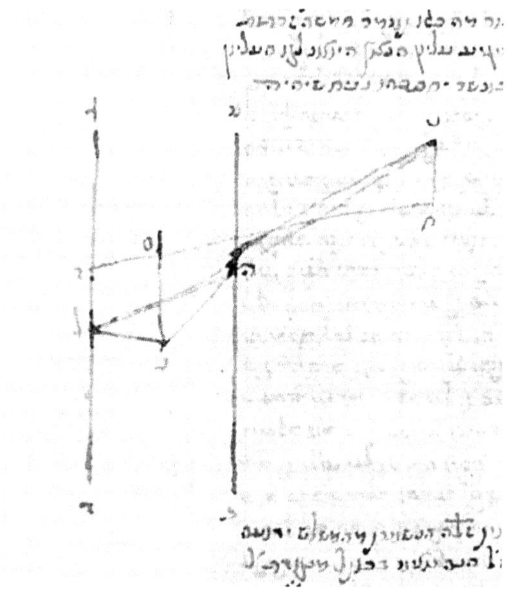

6. Levi ben Gerson (ca. 1340)

Plate 3.1A. The History of the Camera Obscura

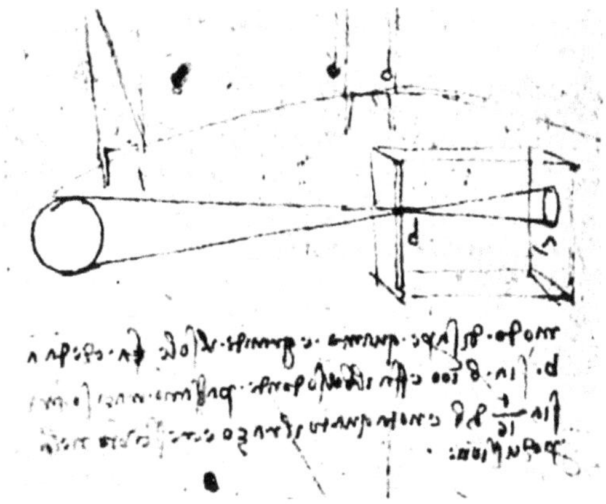

1. Da Vinci (ca. 1490)

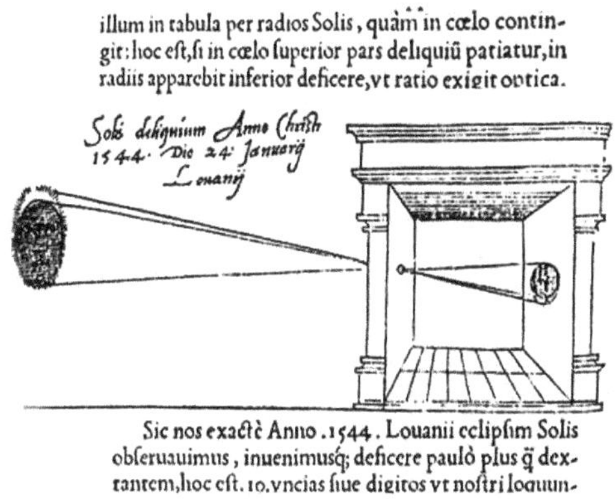

2. Gemma Frisius (1545)

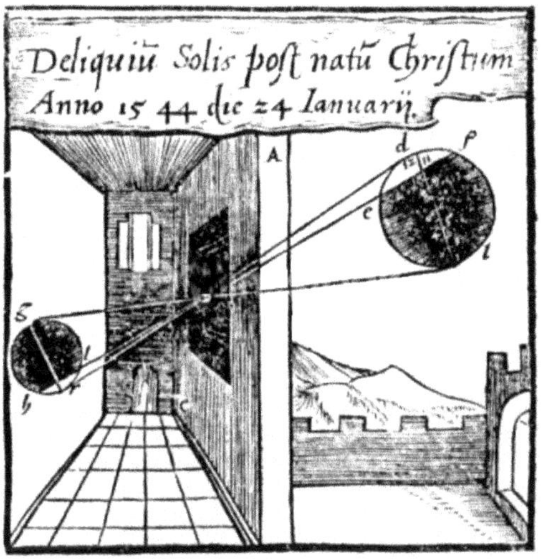

3. Santbech (1561)

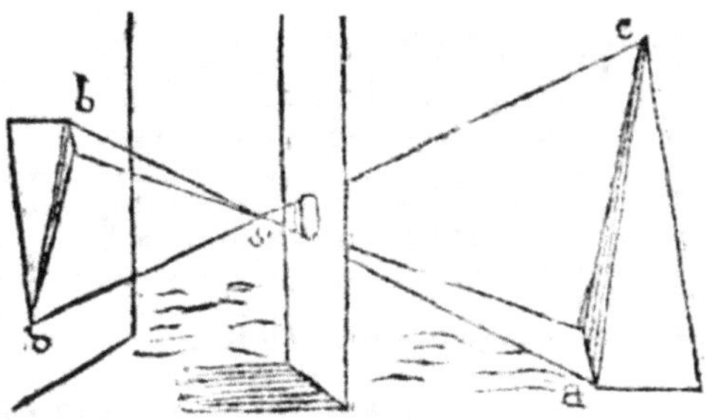

4. Danti (1573)

5. Schwenter (1636)

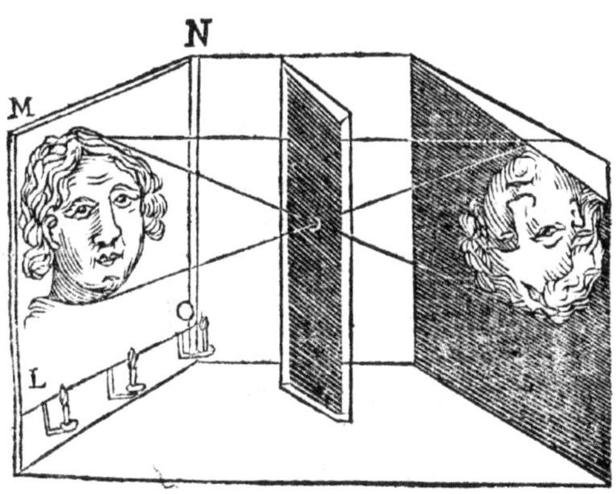

6. Kircher (1646)

Plate 3.1B. The History of the Camera Obscura

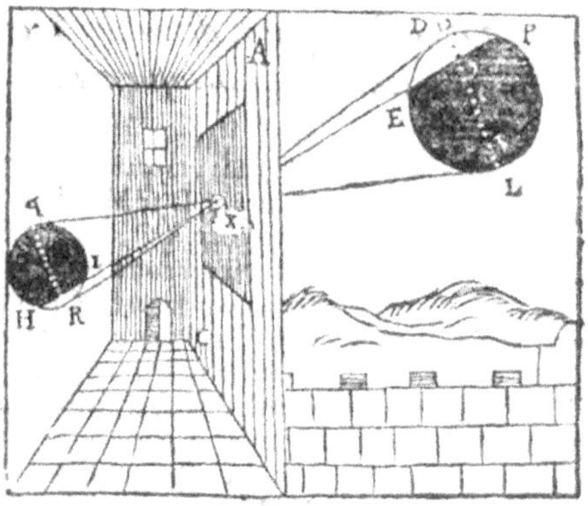

1. Kircher (1646)

2. Kircher (1646)

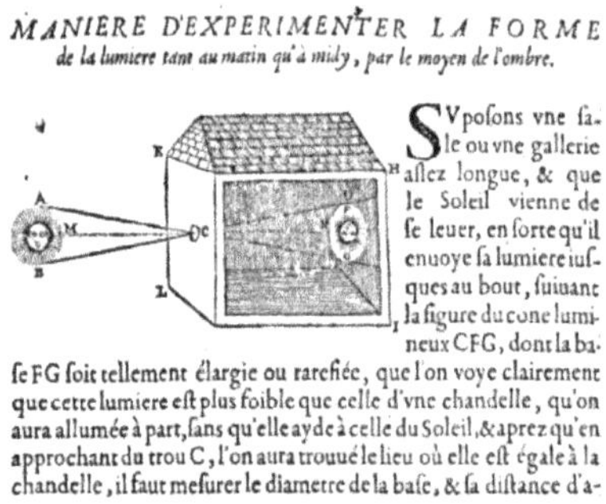

MANIERE D'EXPERIMENTER LA FORME
de la lumiere tant au matin qu'à midy, par le moyen de l'ombre.

SVposons vne sale ou vne gallerie assez longue, & que le Soleil vienne de se leuer, en sorte qu'il enuoye sa lumiere iusques au bout, suiuant la figure du cone lumineux CFG, dont la base FG soit tellement élargie ou rarefiée, que l'on voye clairement que cette lumiere est plus foible que celle d'vne chandelle, qu'on aura allumée à part, sans qu'elle ayde à celle du Soleil, & aprez qu'en approchant du trou C, l'on aura trouué le lieu où elle est égale à la chandelle, il faut mesurer le diametre de la base, & sa distance d'a-

3. Mersenne (1651)

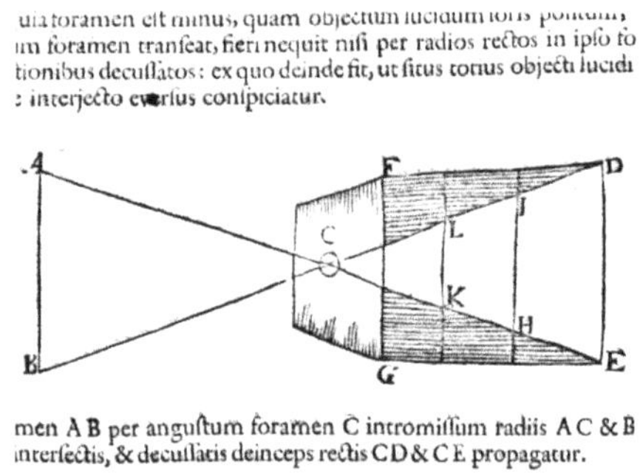

m foramen transeat, fieri nequit nisi per radios rectos in ipso fo
tionibus decussatos: ex quo deinde fit, ut situs totus objecti lucidi
: interjecto eversus conspiciatur.

men A B per angustum foramen C intromissum radiis A C & B interfectis, & decussatis deinceps rectis CD & CE propagatur.

4. Zahn (1685)

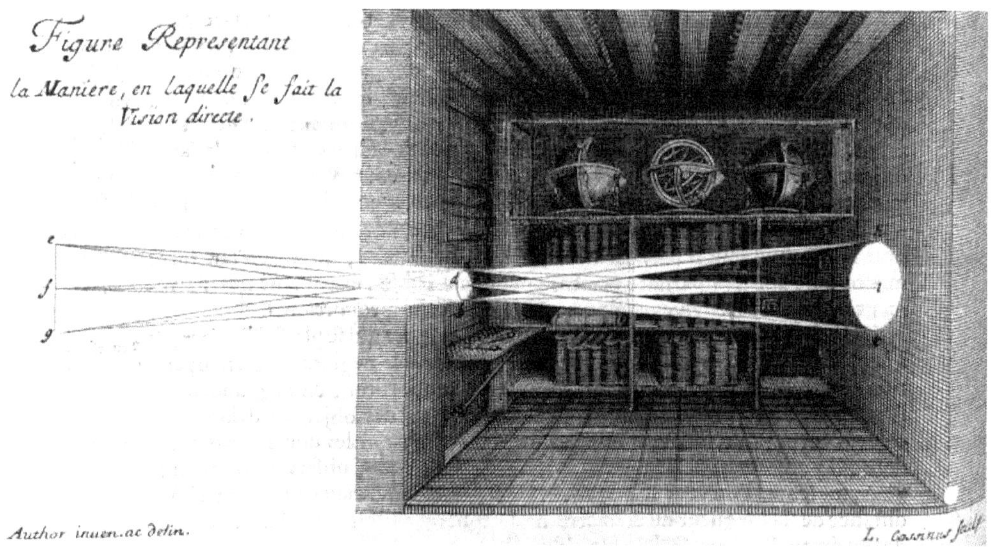

5. Chérubin D'Orléans (1671)

Plate 3.1C. The History of the Camera Obscura

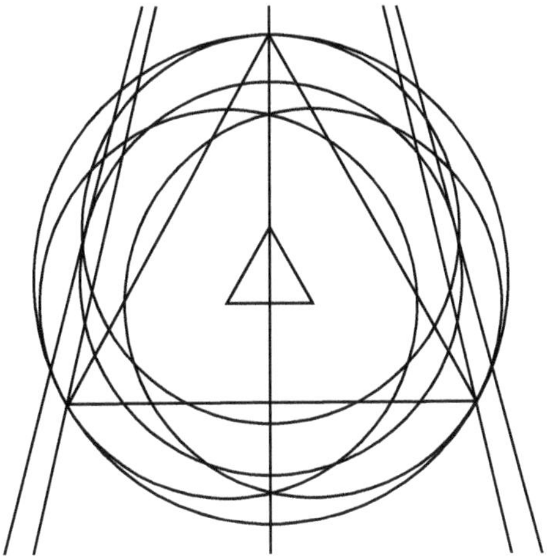

1. Baisiu (ca. 1300)

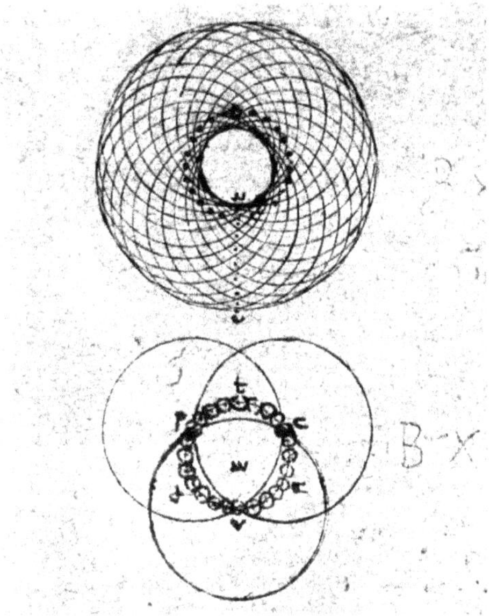

2. Da Vinci (1489)

4. Kepler (July 10, 1600)

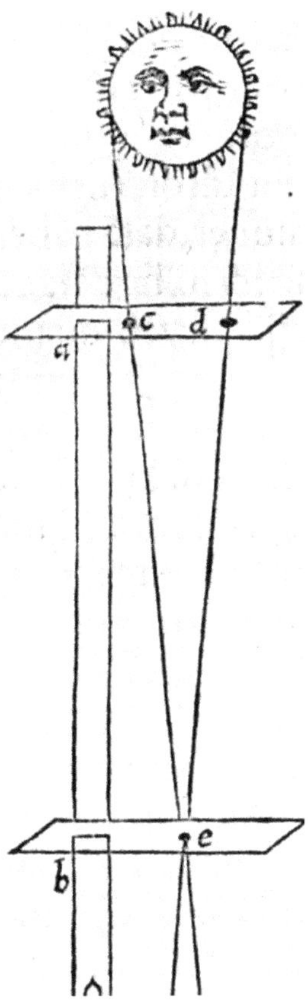

3. Maurolico (1543)

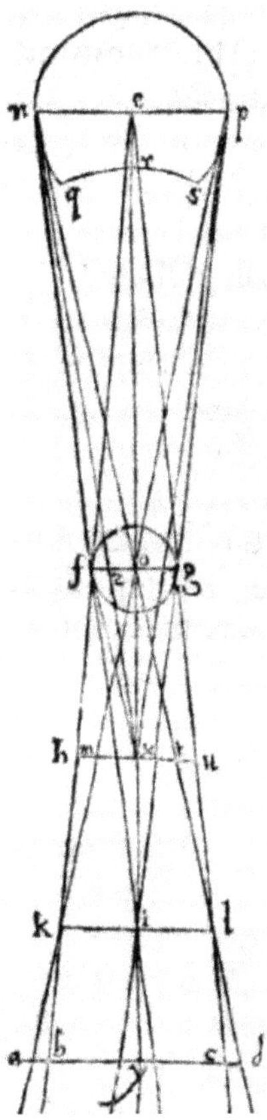

5. Kepler (1604)

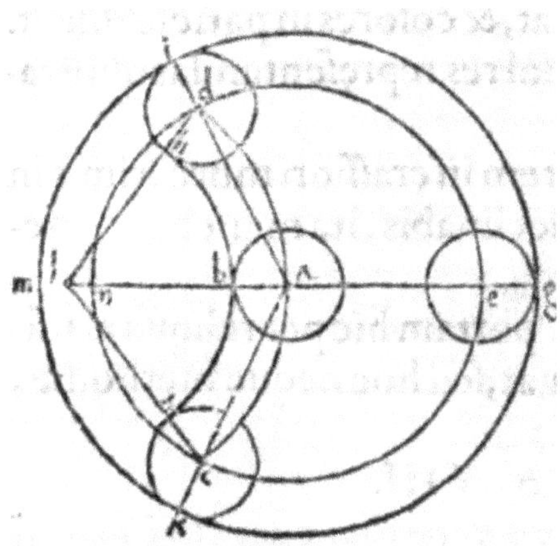

6. Kepler (1604)

Plate 3.2. Explanations of the Formation of the Image

Plate 3.3. The Image through the Aperture Gnomon: in Rayy-Bīrūnī (R1), in Rayy-Marrākushī (R2), and in Samarkand (S)

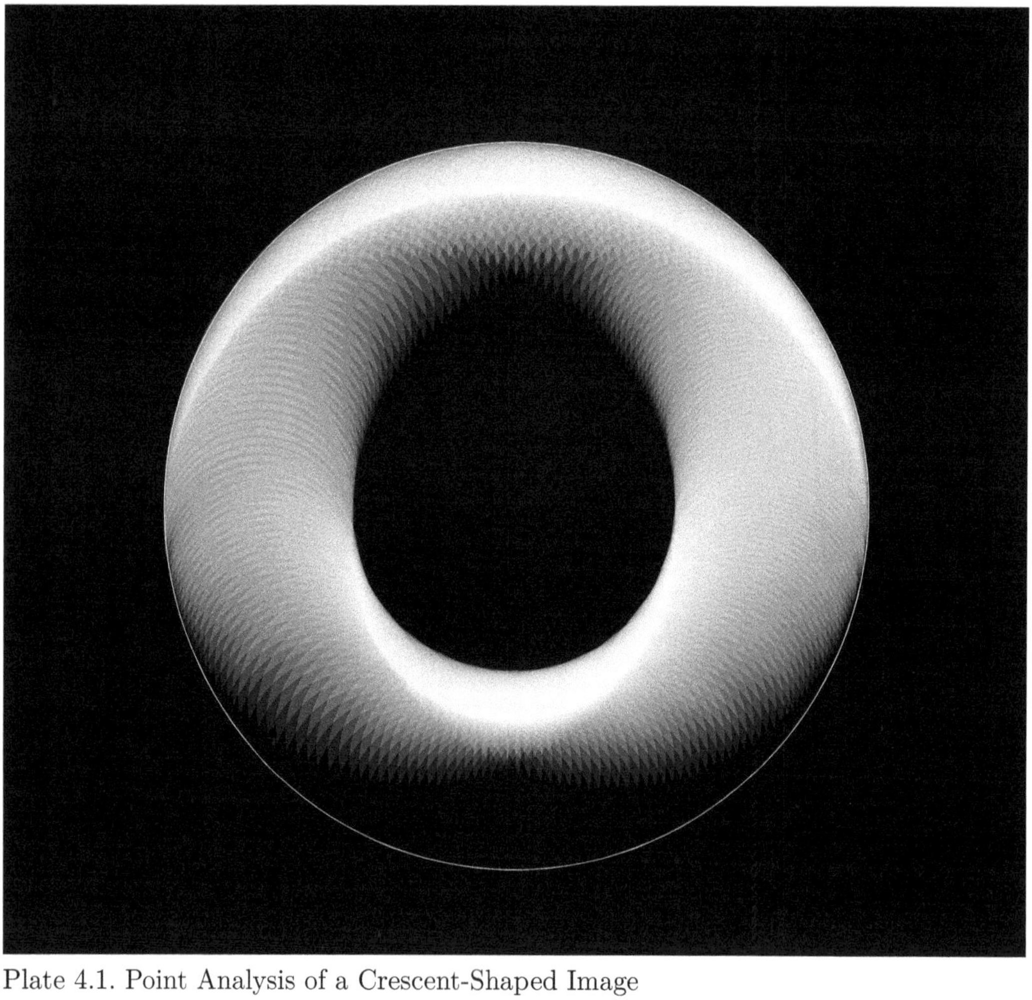

Plate 4.1. Point Analysis of a Crescent-Shaped Image

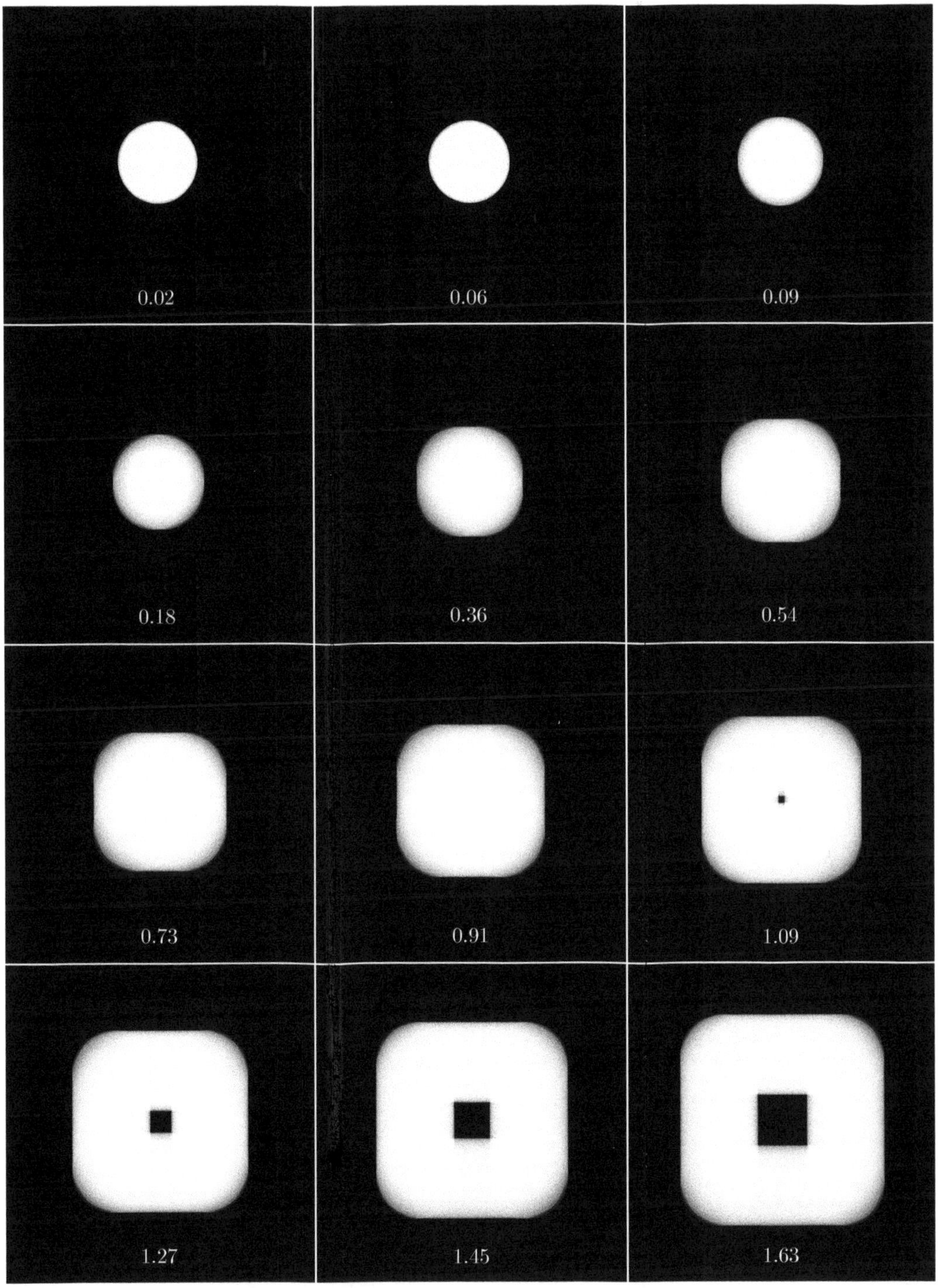

Plate 4.2. The Solar Disc as the Square Aperture Increases

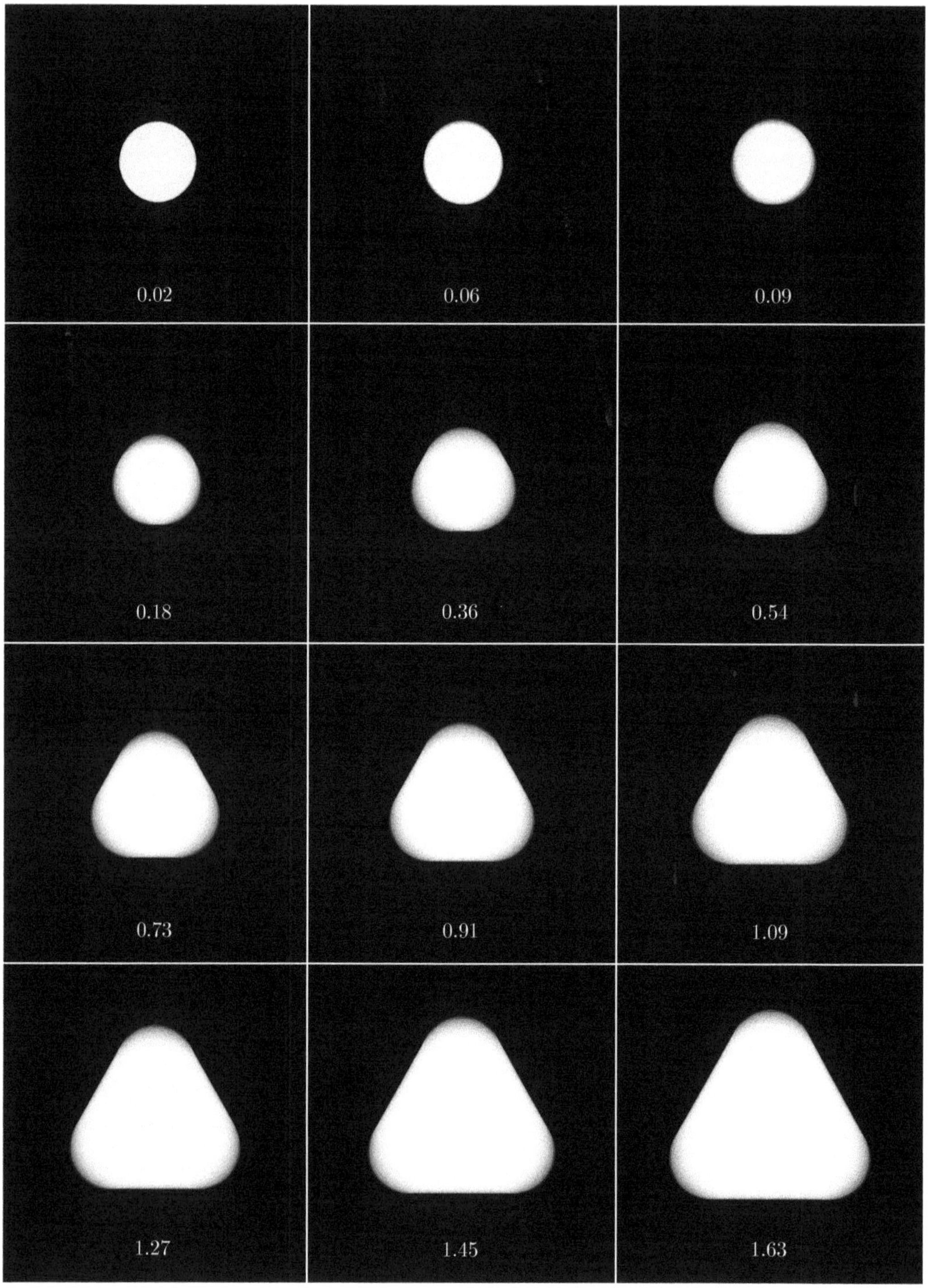

Plate 4.3. The Solar Disc as the Triangular Aperture Increases

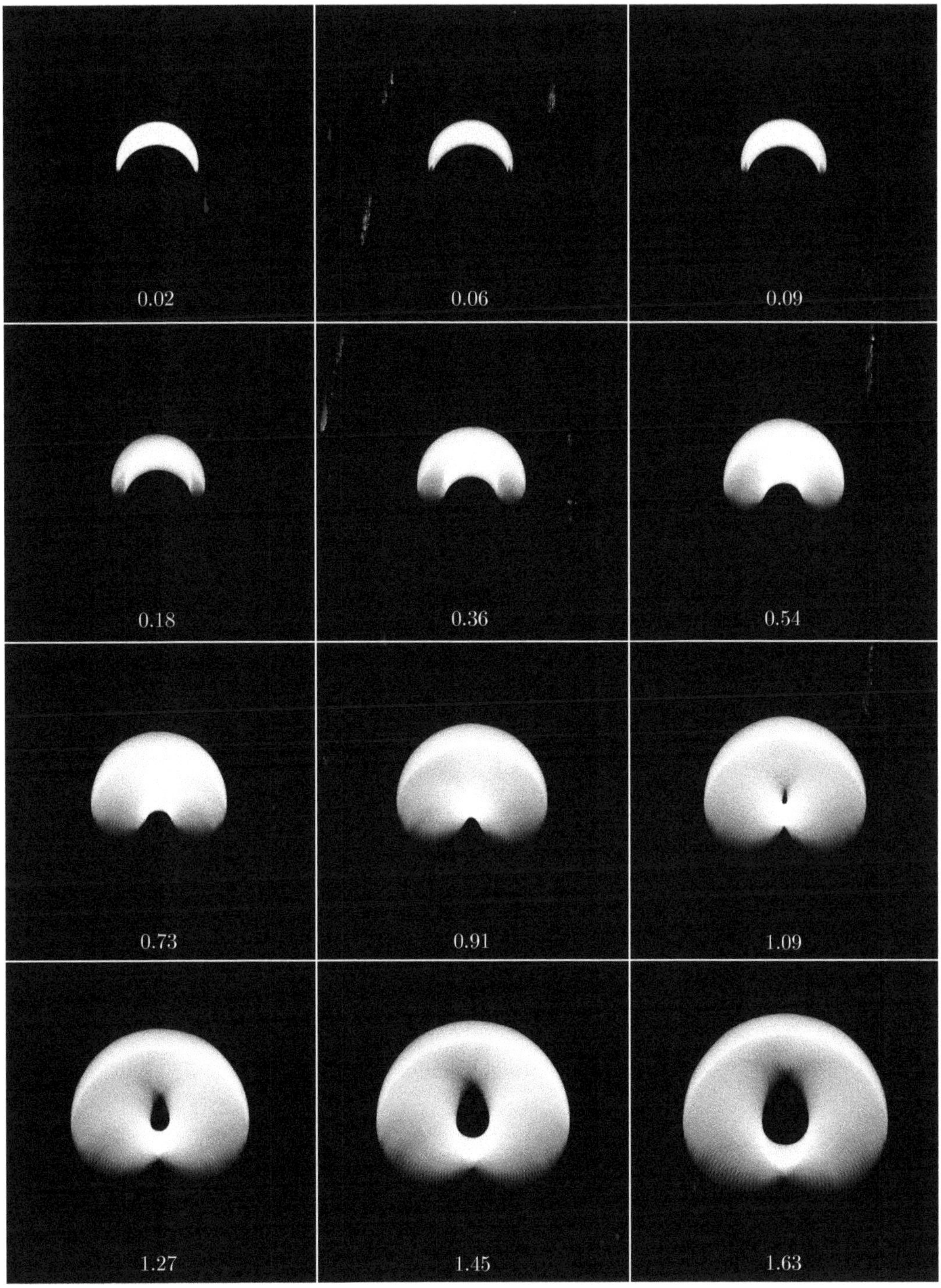

Plate 4.4. The Solar Crescent as the Circular Aperture Increases

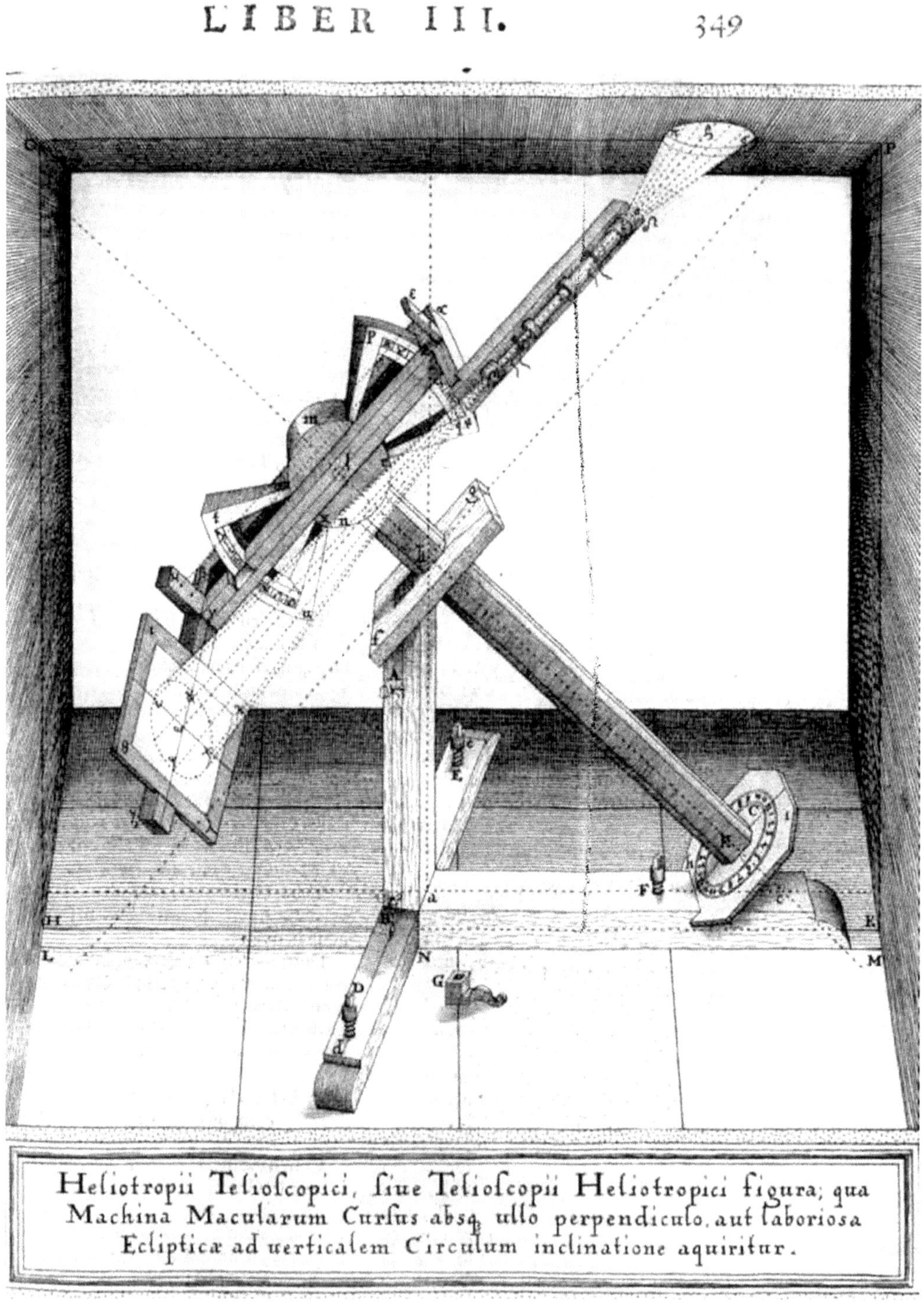

Plate 4.5. Scheiner's Helioscope

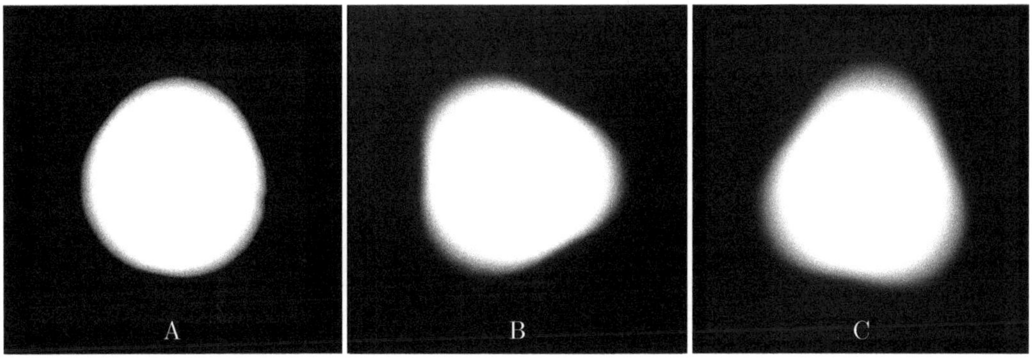

Plate 4.6. Images Obtained through the Helioscope

Plate 4.7. Point Analysis of a Circular Image

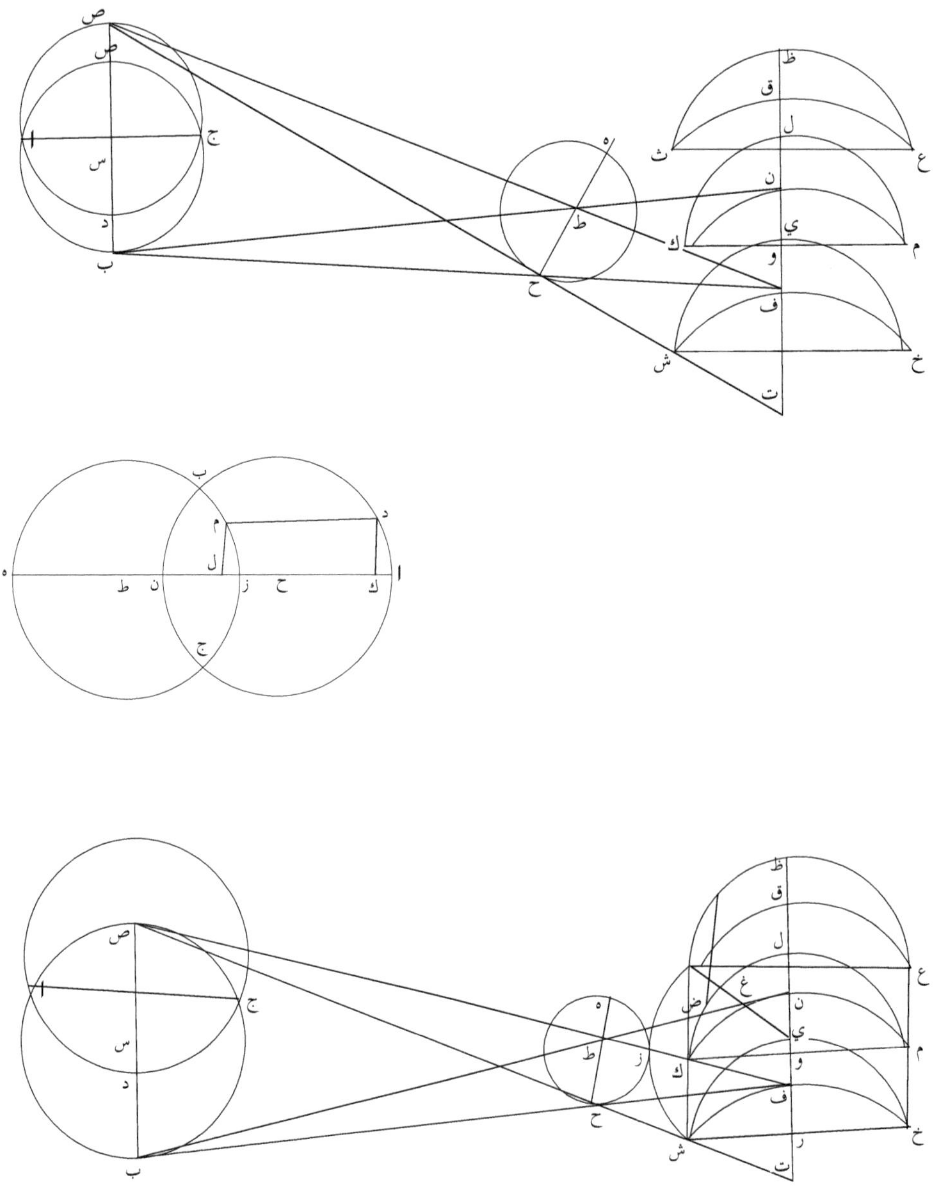

Plate App1. MS F Fātiḥ 3439

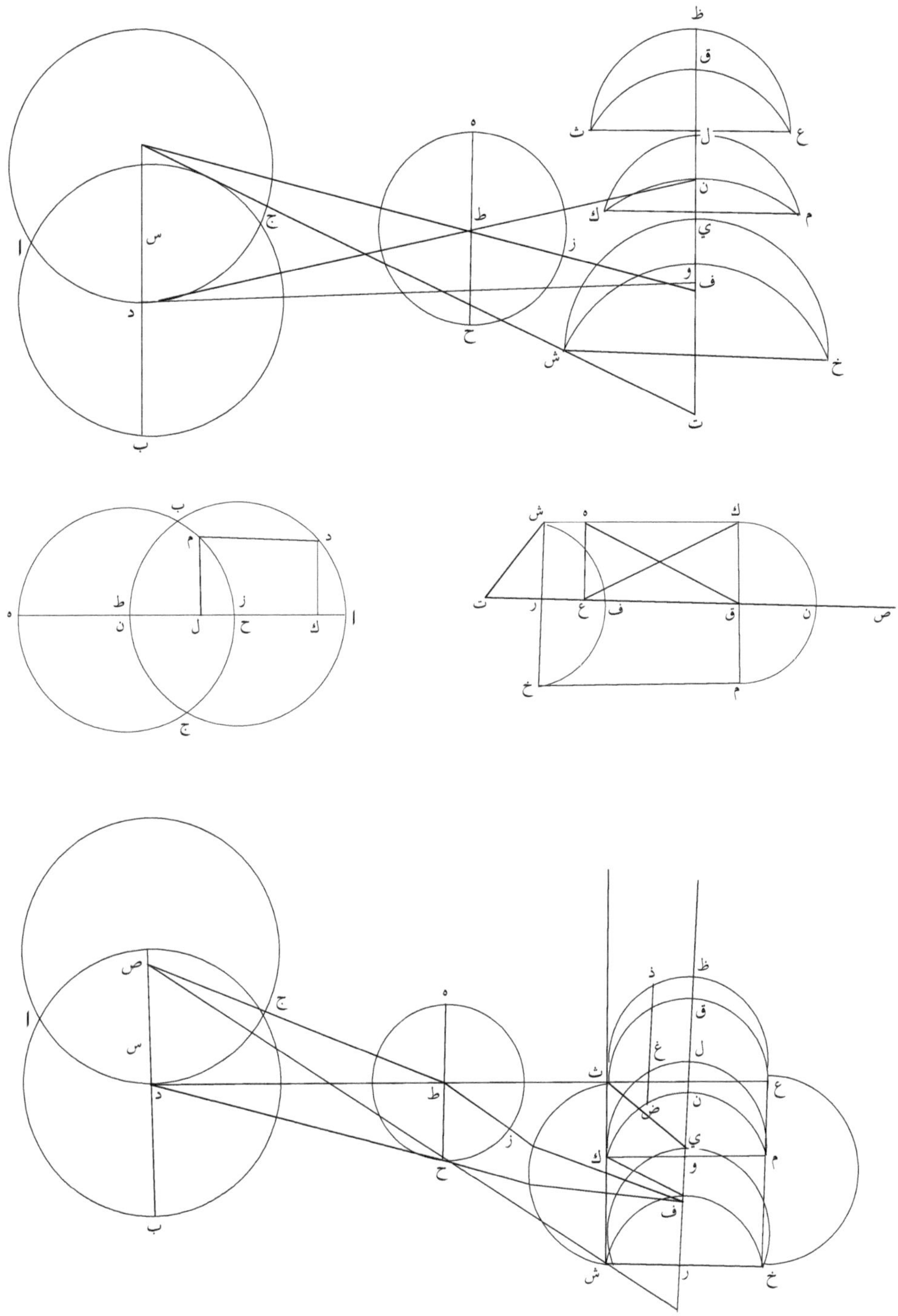

Plate App2. MS B Bodleian Arch. Seld. A32

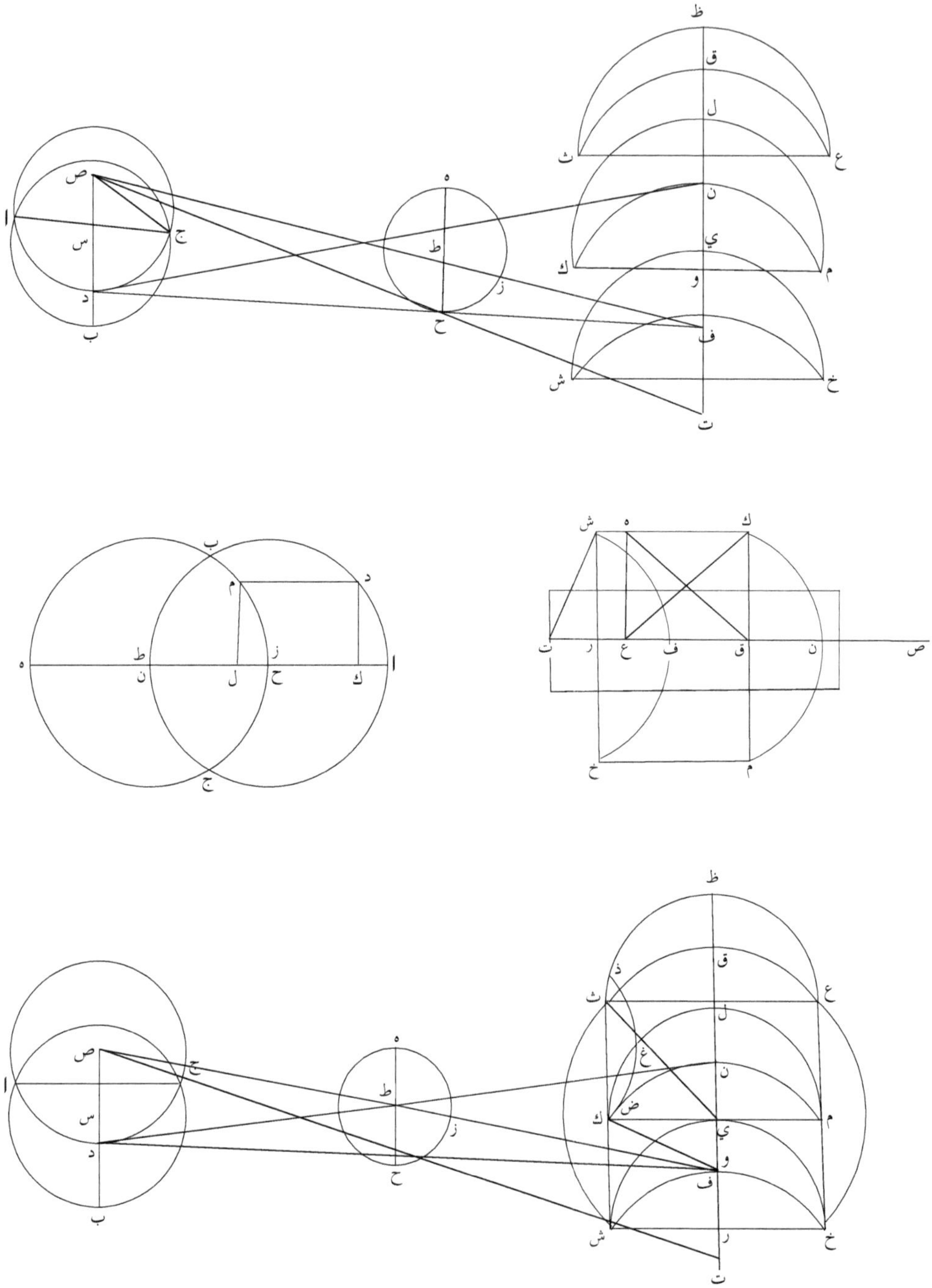

Plate App3. MS P St. Petersburg B1030

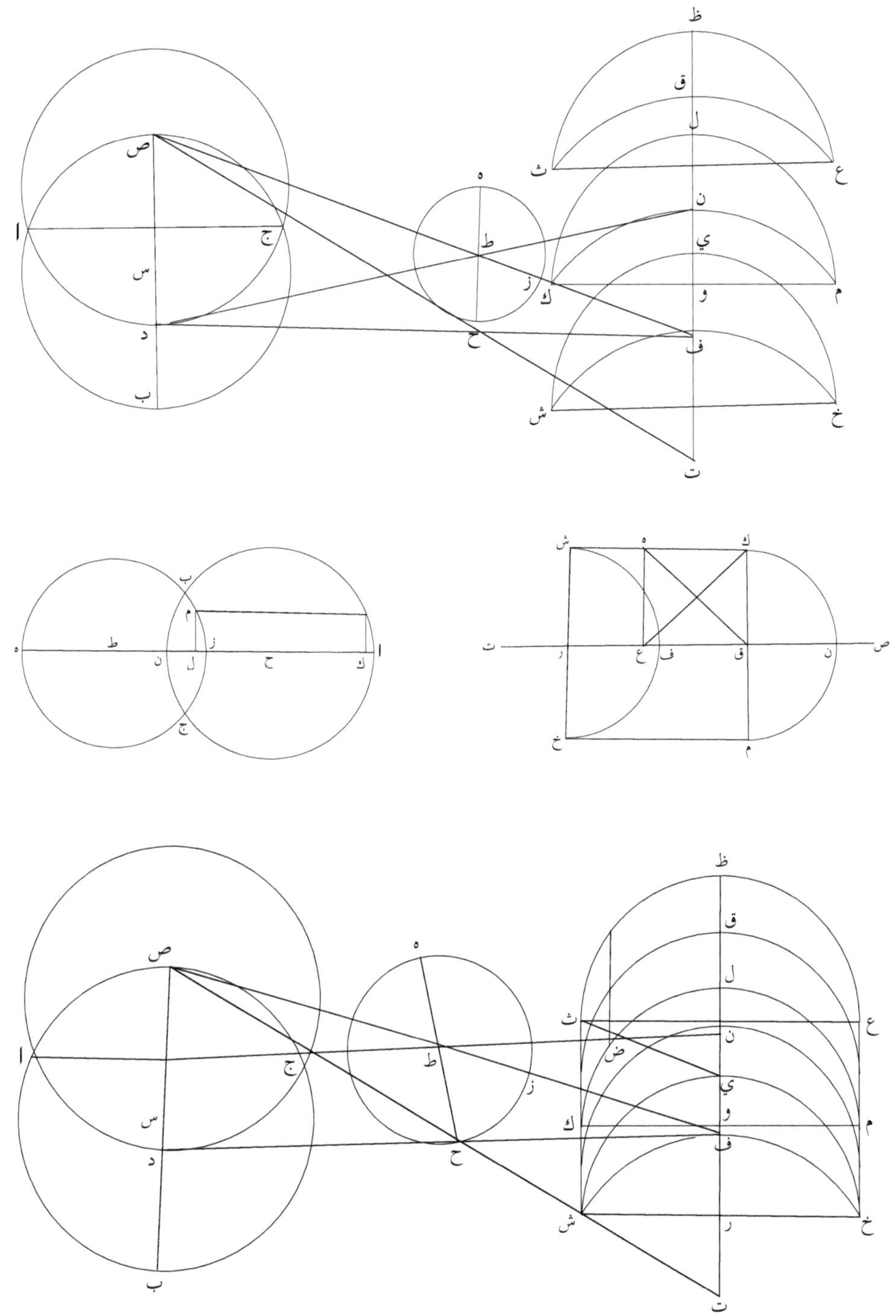

Plate App4. MS O India Office 1270

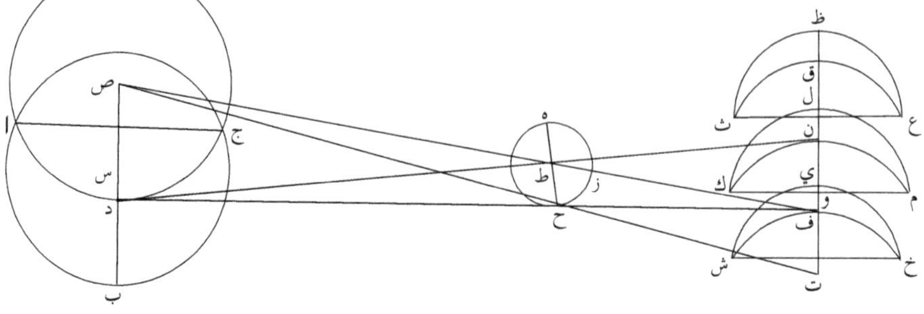

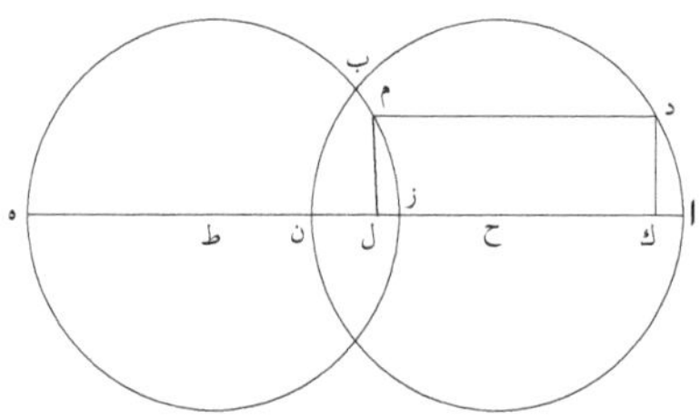

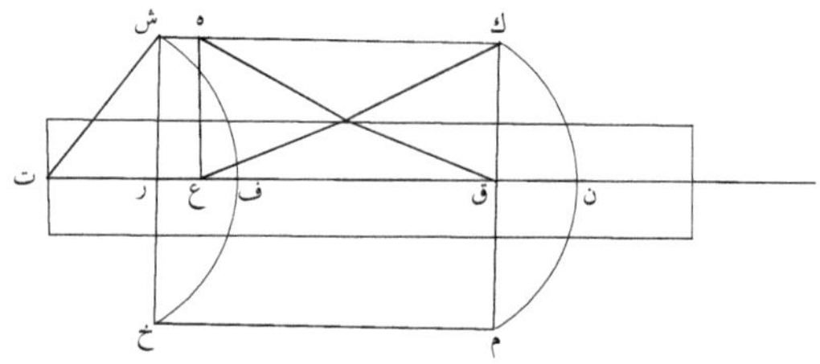

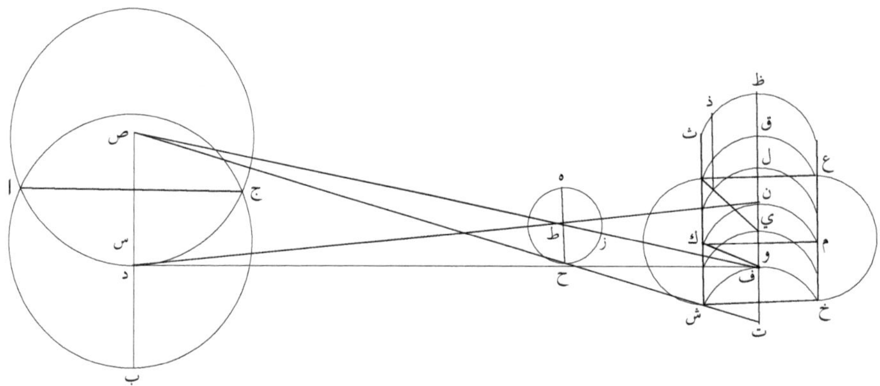

Plate App5. MS L India Office 461

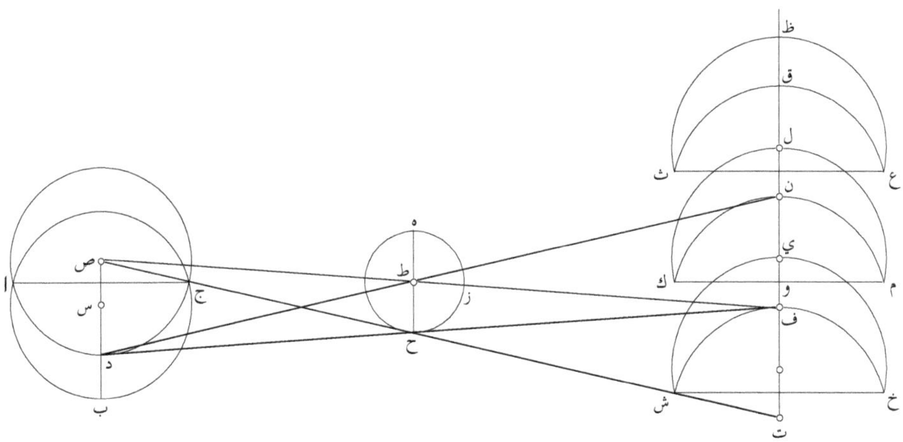

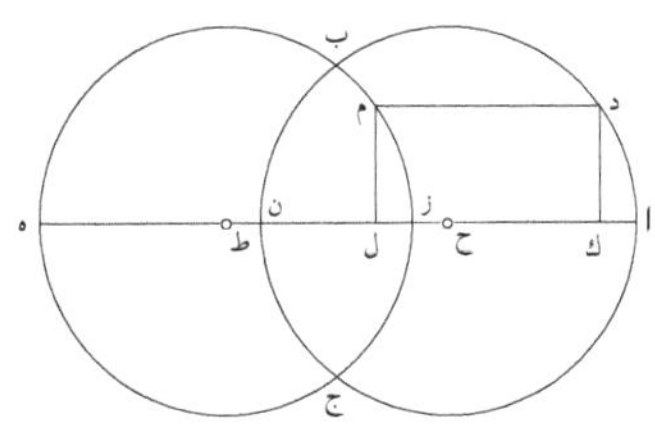

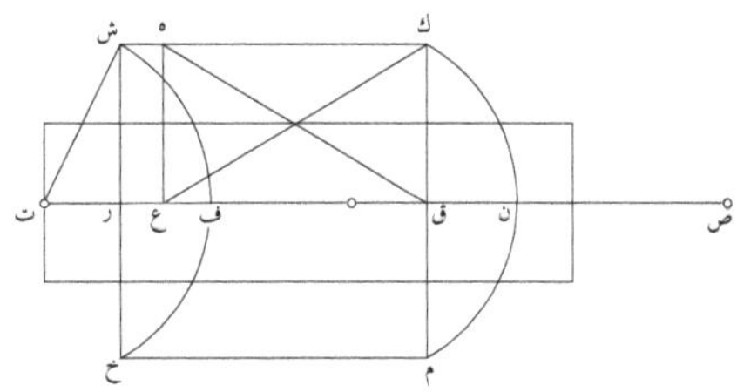

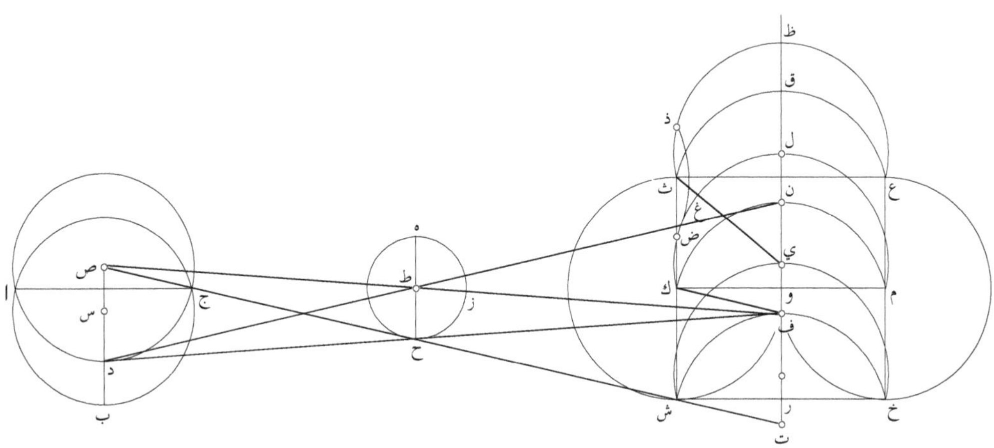

Plate App6. Edited Diagrams

GPSR Compliance
The European Union's (EU) General Product Safety Regulation (GPSR) is a set of rules that requires consumer products to be safe and our obligations to ensure this.

If you have any concerns about our products, you can contact us on

ProductSafety@springernature.com

In case Publisher is established outside the EU, the EU authorized representative is:

Springer Nature Customer Service Center GmbH
Europaplatz 3
69115 Heidelberg, Germany

www.ingramcontent.com/pod-product-compliance
Ingram Content Group UK Ltd.
Pitfield, Milton Keynes, MK11 3LW, UK
UKHW051039091026
15624UKWH00011B/542
* 9 7 8 3 3 1 9 4 7 9 9 0 3 *